Introduction to

# Paint Chemistry

## and Principles of Paint Technology

# Introduction to
# Paint Chemistry
## and Principles of Paint Technology

### G. P. A. Turner
ICI Paints Division, Slough, Berkshire

### THIRD EDITION

London   New York
CHAPMAN AND HALL

First published in 1967 by Chapman and Hall Ltd
11 New Fetter Lane, London EC4P 4EE
Second edition 1980
Third edition 1988
Published in the USA by Chapman and Hall
29 West 35th Street, New York NY 10001

© 1967, 1980, 1988   G. P. A. Turner

Printed in Great Britain by
St Edmundsbury Press Ltd, Bury St Edmunds, Suffolk

ISBN 0 412 29440 0 Hb
ISBN 0 412 29450 8 Pb

British Library Cataloguing in Publication Data

Turner, G. P. A.
  Introduction to paint chemistry and principles of paint
  technology.—3rd ed
  1. Paint
  I. Title
  667'.6        TP936

  ISBN 0-412-29440-0
  ISBN 0-412-29450-8 Pbk

Library of Congress Cataloging-in-Publication Data

Turner, G. P. A. (Gerald Patrick Anthony)
  Introduction to paint chemistry and principles of paint
  technology
  G. P. A. Turner.–3rd ed
  p.    cm.
  Bibliography: p.
  Includes index.
  ISBN 0-412-29440-0.        ISBN 0-412-29450-8 (pbk.)
  1. Paint.     I. Title.
TP935.T8      1988
667'.6–dc19

# Contents

Preface to the third edition            vii
Acknowledgements            ix
Units            xi

**PART ONE:  General science**

1  Atoms to equations      3
2  Inorganic chemistry      17
3  Organic chemistry: paraffins to oils      29
4  Organic chemistry: ethers to isocyanates      44
5  Solid forms      57
6  Colour      70

**PART TWO:  Applied science**

7  Paint: first principles      85
8  Pigmentation      96
9  Solvents      109
10  Paint additives      126
11  Lacquers, emulsion paints and non-aqueous dispersions      138
12  Oil and alkyd paints      154
13  Thermosetting alkyd, polyester and acrylic paints based on nitrogen resins      171
14  Epoxy coatings      183
15  Polyurethanes      198
16  Unsaturated polyesters and acrylics      213
17  Chemical treatment of substrates      228

Appendix: suggestions for further reading      236
Index      239

# Preface to the third edition

*Introduction to Paint Chemistry* was first published in 1967 with the intention of providing both a textbook for students and an introduction to the subject for those with little or no technical knowledge. This remains the objective. The book was completely revised in 1980, but the pace of change continued to quicken. In this third edition, I have sought to bring it up to date with the newest developments in the technology and, with an additional chapter, to emphasize the importance of the painting system as a composite, in which the substrate and its chemistry play a vital role.

The book is divided into two parts. Part One begins at the very basis of matter—its atomic structure—and works step by step through a sufficient selection of chemistry and physics to allow any interested reader to cope with the chemistry and the technology of paint in Part Two. The reader should absorb as much of Part One as he or she feels necessary. It is worth noting, however, that the topics in it are specially selected from a paint point of view and that, for example, detail on oils in Chapter 3, on polymers in Chapter 5 and on light and colour in Chapter 6 could well be missing in some Chemistry degree courses.

Part Two begins with four chapters applicable to paints of every sort and then goes on to six particular paint systems, covering the greater part of paints and varnishes in current use. The classification of paints within these six chapters is largely by drying mechanism. Thus the important family of acrylic finishes does not get a chapter to itself, since the drying mechanisms of the seven types of acrylic coating covered in this book are all different and not essentially acrylic mechanisms. The finishes are described in Chapters 11, 13, 14, 15 and 16. Again, there is no chapter on water-based paints, since water-based paints may be made from a variety of chemically different water-soluble or water-dispersible resins and dry by a variety of mechanisms. The techniques for making resins soluble or dispersible in water are described in Chapters 9 and 11 and exemplified in other chapters.

The Appendix contains suggestions for further reading and I hope that the Index will be sufficiently full to allow quick and easy reference to any topic covered in the book.

G.P.A.T.
1987

# Acknowledgements

I am grateful to the publishers for allowing me this space in which I thank the following for helping me to complete this book:

The Directors of I.C.I., Paints Division, for leave to publish my work.

Mr B. M. Letsky, Industrial Finish Consultant, for the nitrocellulose lacquer formula in Chapter 11.

Rohm & Haas Company, Philadelphia, U.S.A, for the acrylic lacquer formula in Chapter 11.

DSM Resins (UK) Ltd for the metallic car finish formula in Chapter 13.

British Industrial Plastics Ltd for the woodfinish formula in Chapter 13.

Amoco Chemicals UK Ltd for the water-based formula in Chapter 13.

Shell Chemicals for the can coating and electrodeposition primer formulae in Chapter 14.

CIBA-GEIGY (UK) Ltd for the solventless epoxy formula in Chapter 14.

Cargill Blagden for the one pack polyurethane formulae in Chapter 15.

Bayer (UK) Ltd for the two pack polyurethane formula in Chapter 15.

Honeywill and Stein Ltd and U.C.B. s.a. for the U.V. curing acrylic formula in Chapter 16.

Oxford University Press, for permission to base Figs 10 and 11 on an illustration in *Chemical Crystallography*, by C. W. Bunn.

Frederick J. Drake & Co., Chicago, U.S.A, for permission to base Figs 21 and 22 on illustrations in *Color in Decoration and Design*, by F. M. Crewdson.

Reinhold Publishing Corporation, for permission to base Fig. 28 on an illustration in *Principles of Emulsion Technology*, by P. Becher.

The many friends at I.C.I. Paints Division who have helped me with the three editions of this book. To those whose contributions I gratefully acknowledged in the first two editions, I must now add the names of A. J. Backhouse, P. E. T. Baylis, G. J. Clegg, P. S. Collins, J. A. Graystone, D. J. Greenwood, W. Jones, B. C. Joyce, A. J. Naylor, D. J. Walbridge and E. J. West.

Mr G. Light, whose illustrations have so self-evidently stood the test of time.

My mother, Mrs A. M. Best, for once again bearing the brunt of the typing and my family for bearing with me during this revision.
Finally, I acknowledge that all the errors in this book are mine.

G.P.A.T.

# Units

Some of the measurements required to be made for paint and chemical purposes involve units that are new, even to those familiar with the metric system of measurement. For example, how does one describe distances shorter than one millimetre? All the units of measurement to be found in this book are set out below.

## Units of length, area and volume

1 metre (m) = 39·4 inches = 1000 millimetres (mm)
             = 1 million (or $10^6$) microns ($\mu$m)
             = one thousand million (or $10^9$) nanometres (nm)
1 thousandth of an inch (thou or mil) = 25·4 $\mu$m = 25 400 nm
1 litre (l) = 1000 millilitres (ml) = 1·76 pints (Imp)
1 gallon (Imp) = 4544 ml = 1·2 gal (US)

## Units of weight

1 kilogram (kg) = 2·2 pounds = 1000 grams (g)
1 ounce = 28·4 g

## Units of temperature

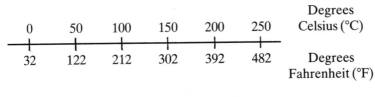

| 0 | 50 | 100 | 150 | 200 | 250 | Degrees Celsius (°C) |
| 32 | 122 | 212 | 302 | 392 | 482 | Degrees Fahrenheit (°F) |

$$x°F = (x - 32) \times \tfrac{5}{9}\,°C$$
$$x°C = \tfrac{9}{5}x + 32\,°F$$

## SI UNITS

Those already familiar with the units of science will be aware that, although the official units of science are now the SI (Système Internationale) units, the older c.g.s. (centimetre, gram, second) units are still widely used. For those interested in such matters, it may be helpful to have some simple conversion aids:

|  | In SI units | In c.g.s. units | Conversion |
|---|---|---|---|
| Basic unit of length | metre (m) | centimetre (cm) | $1\,m = 100\,cm$ |
| Basic unit of mass | kilogram (kg) | gram (g) | $1\,kg = 1000\,g$ |
| Basic unit of time | second (s) | second (s) | — |

Many other units may be derived from these three. Others of interest in this book are:

| | | | |
|---|---|---|---|
| Unit of volume | cubic metre $(m^3)$ | cubic centimetre $(cm^3)$ | $1\,m^3 = 10^6\,cm^3$ |
| | litre (l) | | $1\,l = 1000\,cm^3$ |
| Unit of force | newton (N) | dyne $(gcms^{-2})$ | $1\,N = 10^5\,dynes$ |
| Unit of viscosity | pascal-second (Pa. s) | poise (p) | $1\,Pa.\,s = 10\,p$ |
| Unit of surface tension | newton per m | dyne per cm | $1\,SI\ unit = 10^3$ c.g.s. units |

# PART ONE

# General science

# One

# Atoms to equations

### Atoms and elements

If any pure chemical substance is divided into two pieces and then one piece is again divided into two and the process is repeated so that the size of the piece is continually decreasing, eventually the substance cannot be subdivided any further without being decomposed. At this stage, we have the smallest particle of that substance in existence.

For simple elementary substances the smallest particle is called an *atom*. An atom is very small indeed: a sheet of iron $1\,\mu$m thick would be 4300 atoms thick. These simple substances, which have as their smallest particle an atom, are called *elements*. There are 92 elements found in nature, including such commonplace materials as oxygen and carbon, sulphur and mercury, iron and lead. The *Latin* names of these elements can be abbreviated to one or two letters. These abbreviations are internationally recognized as *symbols* for the elements, e.g. O for oxygen, C for carbon, S for sulphur, Hg for mercury (hydrargyrum), Fe for iron (ferrum) and Pb for lead (plumbum). The 92 elements are listed by their symbols in a table called the Periodic Table (Fig. 1). The elements listed in vertical columns behave similarly in a chemical sense. The elements are given *atomic numbers* from 1 to 92.

The atoms of different elements are different in weight, size and behaviour. Atoms are composed of a number of different electrically charged or neutral particles. An atom consists of a nucleus, which contains relatively heavy particles, some of which are positively charged (*protons*) and some neutral (*neutrons*), and this is surrounded by a number of relatively light negatively charged particles called *electrons*. The electrons are in constant motion about the nucleus and the roughly spherical volume in which they move is the volume of the atom. The electrical charges on protons and electrons are equal, but opposite in sign. The number of electrons in an atom always exactly balances the number of protons, so that an atom is electrically neutral. Atoms of different elements differ in the number of protons and electrons which they contain. The atomic number of an element is the number of protons (or electrons) in one atom of that element.

| GROUP / PERIOD | Ia | IIa | IIIa | IVa | Va | VIa | VIIa | VIII | | | Ib | IIb | IIIb | IVb | Vb | VIb | VIIb | O |
|---|---|---|---|---|---|---|---|---|---|---|---|---|---|---|---|---|---|---|
| 1 | 1 H | | | | | | | | | | | | | | | | 1 H | 2 He |
| 2 | 3 Li | 4 Be | | | | | | | | | | | 5 B | 6 C | 7 N | 8 O | 9 F | 10 Ne |
| 3 | 11 Na | 12 Mg | | | | | | | | | | | 13 Al | 14 Si | 15 P | 16 S | 17 Cl | 18 A |
| 4 | 19 K | 20 Ca | 21 Sc | 22 Ti | 23 V | 24 Cr | 25 Mn | 26 Fe | 27 Co | 28 Ni | 29 Cu | 30 Zn | 31 Ga | 32 Ge | 33 As | 34 Se | 35 Br | 36 Kr |
| 5 | 37 Rb | 38 Sr | 39 Y | 40 Zr | 41 Nb | 42 Mo | 43 Tc | 44 Ru | 45 Rh | 46 Pd | 47 Ag | 48 Cd | 49 In | 50 Sn | 51 Sb | 52 Te | 53 I | 54 Xe |
| 6 | 55 Cs | 56 Ba | 57 La ■ | 72 Hf | 73 Ta | 74 W | 75 Re | 76 Os | 77 Ir | 78 Pt | 79 Au | 80 Hg | 81 Tl | 82 Pb | 83 Bi | 84 Po | 85 At | 86 Rn |
| 7 | 87 Fr | 88 Ra | 89 Ac ■■ | | | | | | | | | | | | | | | |

| LANTHANIDE SERIES ■ | 58 Ce | 59 Pr | 60 Nd | 61 Pm | 62 Sm | 63 Eu | 64 Gd | 65 Tb | 66 Dy | 67 Ho | 68 Er | 69 Tm | 70 Yb | 71 Lu |
|---|---|---|---|---|---|---|---|---|---|---|---|---|---|---|

| ACTINIDE SERIES ■■ | 90 Th | 91 Pa | 92 U |
|---|---|---|---|

**Fig. 1** The Periodic Table.

| Atomic no. | Symbol | Element | Atomic no. | Symbol | Element | Atomic no. | Symbol | Element |
|---|---|---|---|---|---|---|---|---|
| 1 | H | Hydrogen | 32 | Ge | Germanium | 63 | Eu | Europium |
| 2 | He | Helium | 33 | As | Arsenic | 64 | Gd | Gadolinium |
| 3 | Li | Lithium | 34 | Se | Selenium | 65 | Tb | Terbium |
| 4 | Be | Beryllium | 35 | Br | Bromine | 66 | Dy | Dysprosium |
| 5 | B | Boron | 36 | Kr | Krypton | 67 | Ho | Holmium |
| 6 | C | Carbon | 37 | Rb | Rubidium | 68 | Er | Erbium |
| 7 | N | Nitrogen | 38 | Sr | Strontium | 69 | Tm | Thulium |
| 8 | O | Oxygen | 39 | Yt | Yttrium | 70 | Yb | Ytterbium |
| 9 | F | Fluorine | 40 | Zr | Zirconium | 71 | Lu | Lutecium |
| 10 | Ne | Neon | 41 | Nb | Niobium | 72 | Hf | Hafnium |
| 11 | Na | Sodium | 42 | Mo | Molybdenum | 73 | Ta | Tantalum |
| 12 | Mg | Magnesium | 43 | Tc | Technetium | 74 | W | Tungsten |
| 13 | Al | Aluminium | 44 | Ru | Ruthenium | 75 | Re | Rhenium |
| 14 | Si | Silicon | 45 | Rh | Rhodium | 76 | Os | Osmium |
| 15 | P | Phosphorus | 46 | Pd | Palladium | 77 | Ir | Iridium |
| 16 | S | Sulphur | 47 | Ag | Silver | 78 | Pt | Platinum |
| 17 | Cl | Chlorine | 48 | Cd | Cadmium | 79 | Au | Gold |
| 18 | A | Argon | 49 | In | Indium | 80 | Hg | Mercury |
| 19 | K | Potassium | 50 | Sn | Tin | 81 | Tl | Thallium |
| 20 | Ca | Calcium | 51 | Sb | Antimony | 82 | Pb | Lead |
| 21 | Sc | Scandium | 52 | Te | Tellurium | 83 | Bi | Bismuth |
| 22 | Ti | Titanium | 53 | I | Iodine | 84 | Po | Polonium |
| 23 | V | Vanadium | 54 | Xe | Xenon | 85 | At | Astatine |
| 24 | Cr | Chromium | 55 | Cs | Caesium | 86 | Rn | Radon |
| 25 | Mn | Manganese | 56 | Ba | Barium | 87 | Fr | Francium |
| 26 | Fe | Iron | 57 | La | Lanthanum | 88 | Ra | Radium |
| 27 | Co | Cobalt | 58 | Ce | Cerium | 89 | Ac | Actinium |
| 28 | Ni | Nickel | 59 | Pr | Praesodymium | 90 | Th | Thorium |
| 29 | Cu | Copper | 60 | Nd | Neodymium | 91 | Pa | Protoactinium |
| 30 | Zn | Zinc | 61 | Pm | Promethium | 92 | U | Uranium |
| 31 | Ga | Gallium | 62 | Sm | Samarium | | | |

Everyone knows that it is now possible to split an atom, but since the products contain less protons and neutrons than the original atom, they are therefore atoms of *different* (lighter) elements. Thus it is still true that the smallest particle of an element is an atom of that element: anything smaller does not possess the properties of that element.

## Molecules and compounds

All other substances are compounded from the elements. These substances, known as *compounds,* have as their smallest particle, a *molecule.* A molecule consists of two or more atoms held together by chemical bonds (see below). A molecule of a compound must contain atoms of at least two different elements.

Compounds, like elements, can be represented by symbols. These symbols reveal the proportions of the atoms of different elements present in the molecule. Thus water, $H_2O$, contains two atoms of hydrogen and one of oxygen per molecule. This symbol is also described as the chemical *formula* of water.

It is customary in most chemical formulae of inorganic compounds (Chapter 2), to write metallic atoms first (if there are any), e.g. NaCl, sodium chloride, or common salt. In the Periodic Table, the metallic elements extend from the left almost three-quarters of the way across the table: all the A groups, groups VIII, IB, IIB and the lower parts of groups IIIB–VIB. In spite of this, the non-metallic elements have the larger role to play in chemistry, as we shall see (Chapters 3–16).

A *mixture* of elements, without chemical bonding between the different types of atoms, is not a compound. Mixtures of elements can often be separated by simple physical means, e.g. a magnet removes iron from a mixture of iron filings and sulphur. But the main distinction between mixtures and compounds is that the proportions of the elements in a mixture may vary considerably from sample to sample: in a given compound the proportions are always the same.

## The states of matter

The knowledge that all matter is composed of atoms and molecules permits an explanation of the physical properties of matter. At any particular temperature any substance exists as either a solid, a liquid, or a gas. If the temperature is varied, the same substance may pass through all states, provided that it remains stable. The changes are best followed on a molecular scale.

Identical molecules possess forces of attraction for one another at close range. At the same time, all molecules at temperatures above the absolute zero (approx. $-273\,°C$) possess energy, received from their surroundings in the form of heat, light or radiations, or transferred from other molecules in the course of molecular collisions. The molecules are in a constant state of motion as a result of the energy they possess. These facts explain the states of matter.

## The solid state

In the solid state the forces of attraction are strong enough to over-come the motion of the molecules, so that they are closely packed in an ordered arrangement. Nevertheless they are moving by vibrating backwards and forwards (Fig. 2). As the solid is heated, the vibration becomes more vigorous, until finally the molecules have sufficient energy to overcome the attractive forces. As they gain freedom of movement, the rigid pattern of atoms is broken and the structure collapses. The solid has melted. Melting occurs at a fixed temperature called the *melting point* and a fixed amount of heat, known as the *latent heat of fusion,* is required to melt each gram of the pure solid compound.

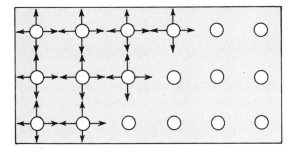

**Fig. 2** The arrangement of molecules in a solid.

## The liquid state

Although the molecules can now move about, they are still close to one another and the attractive forces prevent complete escape. The molecules at the surface of any volume of liquid are in a special position. They are not attracted by neighbours in all directions, since on one side their only neighbours are occasional gas molecules in the air. There is little pull in this direction: the attractive forces tend to draw surface molecules together and in towards the centre of the liquid. The liquid behaves as if it had a skin and the surface molecules are described as being under a *surface tension* (Fig. 3). Surface tension is a force usually expressed in dynes per cm (newtons per metre in SI units gives figures one thousandth smaller and are not widely used). Surface tensions of simple liquids vary from 73 for water through 30 for xylene and 25 for methyl ethyl ketone to 18 for *n*–hexane (see Chapters 3, 4 and 9).

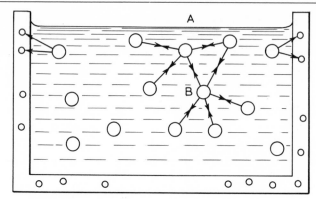

**Fig. 3** The arrangement of molecules in a liquid; surface tension.
A. Surface molecules attracted to centre of liquid.
B. Interior molecules attracted in all directions.

*Viscosity* is the resistance, drag or friction on any object moving through a liquid, or the resistance (due to the chemical attraction and physical shape of molecules) which hinders the flow of the liquid. Imagine pouring a collection of similarly shaped stones from a sack. They jostle and drag against one another: there is physical and mechanical resistance to flow. If we also imagine each stone as a small magnet and further, as having a built-in motor driving it along in a straight line, then we are nearer to the conditions operating inside a liquid. There the movement of one molecule in a given direction is hindered by (*a*) collisions with other molecules (*b*) physical entanglement with the other molecules, due to irregularities of shape and (*c*) attraction for other molecules in the immediate vicinity. The result of all this is viscosity.

The scientific measure of viscosity, the *coefficient of viscosity* of a liquid, $\eta$ (pronounced 'eta'), is given by Newton's equation

$$\eta = \frac{\text{stress}}{\text{rate of shear}}$$

The *stress* is the frictional force operating per unit area, while the *rate of shear* is the variation in velocity of the moving liquid with depth of liquid. If we imagine a flowing river, at any instant the water moving fastest is at the surface, while the water in contact with the river bed is not moving at all. Considering the layers in between, the nearer we get to the top, the faster the water is flowing.

Since

$$\text{stress} = \frac{\text{force}}{\text{area}} = \frac{\text{mass} \times \text{acceleration}}{\text{length}^2} = \frac{\text{mass} \times (\text{length}/\text{time}^2)}{\text{length}^2}$$

$$= \frac{\text{mass}}{\text{length} \times \text{time}^2} \left( \text{or, in S.I. units,} \quad \frac{\text{kg}}{\text{m} \times \text{s}^2} \right)$$

and

$$\text{rate of shear} = \frac{\text{velocity}}{\text{length}} = \frac{\text{length/time}}{\text{length}} = \frac{1}{\text{time}} \left( \text{or } \frac{1}{\text{s}} \text{ in S.I. units} \right)$$

then $\eta$ is measured in $\dfrac{\text{kg}}{\text{m} \times \text{s}^2} \times \text{s} = \dfrac{\text{kg}}{\text{m} \times \text{s}}$

It is more usual to express $\eta$ in units one-tenth of this size and these units are called *poises* after the physicist, Poiseuille. In scientific terms, we therefore speak of the coefficient of viscosity, or simply the viscosity, of a liquid as being so many poises.

The viscosity of a liquid varies appreciably with small temperature changes. A temperature rise provides the molecules with more energy, so that they move faster. They are thus more capable of overcoming the attractive forces and getting away from their neighbours. This means that the liquid will flow more readily and the viscosity will drop.

As the liquid is heated the energy is shared out unequally between the molecules, so that, at the surface, some are eventually moving fast enough to escape the attractions of their neighbours and fly off into the air. This process is known as *evaporation*. Finally, all the surface molecules escape readily and boiling takes place. The temperature of the liquid remains steady at the *boiling point,* until all the liquid has vapourized. A fixed amount of heat, known as the *latent heat of evaporation,* is required to boil each gram of pure liquid.

**The gaseous state**

The molecules in a gas are widely spaced and the forces of attraction come into play only when chance encounters are made by the moving molecules. Between collisions they travel in straight lines at several hundred miles an hour. The movements of the molecules are limited by the walls of the container in which the gas is placed and the molecules exert *pressure* on those walls by their repeated impacts on them. The higher the temperature, the more energy the molecules possess, the faster they move and the greater the impact. Hence gas pressure increases with temperature.

If the gas is cooled, the molecules get slower and slower until eventually they cannot escape from one another after chance encounters. The attractive forces prevent this. When sufficient molecules are in close proximity they form a drop of liquid and the gas is said to have condensed. *Condensation* takes place at or below the boiling point. Further cooling leads to further slowing down of the molecules, so that the attractive forces re-form the rigid pattern of molecules of the solid state. Solidification takes place at the melting (or freezing) point.

## Solutions, suspensions and colloids

If a solid and liquid are mixed, one of three things can happen:

(1) If the attraction between liquid and solid molecules is less than that between solid molecules, the liquid will do no more than separate solid *particles* from one another. These particles, each composed of hundreds, thousands or millions of molecules, will be dispersed throughout the liquid on shaking. The result is a *suspension* of solid particles, which will ultimately settle out on standing.

(2) If there is a strong attraction between liquid and solid molecules, liquid molecules will penetrate the solid structure and surface solid molecules will break away and mingle with the liquid molecules. Gradually the solid structure is eroded away until it is no longer visible, because the liquid and solid molecules are uniformly mixed. The solid had dissolved and a *solution* has formed.

(3) A chemical reaction may occur. The products of this – if a liquid and a solid – will behave as in (1) and (2).

### Solutions

There is a limit to the weight of crystalline solid that will dissolve in 100 g of liquid. This quantity in grams is known as the *solubility* of the solid. Solubility varies with temperature, usually increasing with a rise in temperature. A solution containing the maximum amount of dissolved solid is a *saturated solution*. The liquid in any solution is known as the *solvent* and the solid as the *solute*.

### Colloidal suspensions

A suspension may contain particles too small to be distinguished by the naked eye. These particles are larger than large molecules, but can be seen with the aid of a very powerful microscope (magnification $\times 900$), only if the particle diameter is $0 \cdot 4 \mu$m or more. Smaller particles are not directly visible, because $0 \cdot 4 \mu$m is the lower wavelength limit of the visible spectrum (see Chapter 6), but they can be seen with the ultramicroscope or with the electron microscope. As the particles settle out, they are constantly bombarded by the liquid molecules, which are in continual motion. Such collisions do not affect large particles, but the small particles rebound slightly and the frequency of the collisions causes them to alter course and follow a zigzag path. This is known as *Brownian movement*. Such random movement delays settlement. The particle surfaces may also bear an electric charge. When two particles come close together, the like charges repel one another. This also delays settlement. Such fine particles are said to be in the *colloidal* state and colloidal suspensions can have extreme stability and resist settling out for very long periods.

A colloidal suspension of high solid content may be opaque (e.g. pigment in paint), but if the solid content is low it may appear transparent, like a solution. However, light entering the colloidal dispersion from the side will be scattered and the eye can detect the particles as scintillations of scattered light. This is called the *Tyndall effect.*

Another form of colloidal dispersion can be produced from two liquids that will not mix. Agitation will disperse one of them as fine droplets in the other. The product is an *emulsion.* The emulsion can be very stable if the dispersed droplets are electrically charged or have adsorbed a surfactant (Chapter 10) at their surfaces. Milk is an emulsion of fat in water.

## Valency and chemical bonds

We have been discussing the behaviour of molecules during certain physical changes, but now we shall consider the internal structure of molecules in more detail. How are the atoms bonded together and what determines the proportions of the different kinds of atoms?

The proportions are determined by the valencies of the elements concerned. The *valency* of an element is the number of chemical bonds that one atom of an element can make. If two atoms come so close together that the spaces in which their electrons move overlap, then at that overlap the negative electrons are attracted by both positive atomic nuclei. Under these circumstances a pair of electrons become 'fixed' at an equilibrium position between the two nuclei. The atoms are not easily separated: a *chemical bond* has formed between them. In a *covalent bond* one electron is supplied by each atom. In a *co-ordinate bond* both electrons are provided by the same atom. The electrons concerned are known as the *valency electrons* and there is a fixed number (less than eight) in each atom. The number varies from element to element. For some elements the valency is equal to the number of valency electrons, i.e. the number of electrons available for 'pairing up' in chemical bonds. This number is the Periodic Table Group Number of the element. For others, the valency is decided by the number of 'spaces' available for bonding electrons from other atoms (eight minus the Group Number). These elements have fixed valencies; some of the more common elements in these categories are:

| Elements donating valency electrons | Valency | Elements accepting valency electrons |
|---|---|---|
| H, Na, K, Ag | 1 | H, F, Cl, Br, I |
| Mg, Ca, Ba, Zn | 2 | O |
| B, Al | 3 | |
| C, Si | 4 | C, Si |

Other elements (e.g. P,N and S) sometimes have the valency permitted by the number of valency electrons and at other times, the valency permitted by the number of 'spaces'. Elements in groups IVA to IIB inclusive and a few others, have atomic structures which permit an increase in the number of valency electrons under some circumstances. Therefore these elements also possess more than one valency. The two types are classified together as 'elements of variable valency'.

| Valency | Elements | Names of valency states |
|---------|----------|-------------------------|
| 1 or 2 | Cu | Cuprous (1), Cupric (2) |
|  | Hg | Mercurous (1), Mercuric (2) |
| 1 or 3 | Au | Aurous (1), Auric (3) |
| 2 or 3 | Fe | Ferrous (2), Ferric (3) |
|  | Co | Cobaltous (2), Cobaltic (3) |
| 2 or 4 | Sn | Stannous (2), Stannic (4) |
|  | Pb | Plumbous (2), Plumbic (4) |
|  | Ni | Nickelous (2), Nickelic (4) |
| 2 or 6 | S | — |
| 3 or 5 | N | — |
|  | P | — |
| 2, 3 or 6 | Cr | Chromous (2), Chromic (3) |
| 2, 4 or 7 | Mn | Manganous (2) |

Note that the *-ous* ending is used to describe the lower valency and *-ic* the higher valency in each case.

Elements of variable valency usually exhibit their lowest valency, but will change valency under the right chemical conditions (see *Oxidation*, Chapter 2).

An element always uses its full valency to form chemical bonds. Should one of these bonds break in the course of chemical reaction (see below), the element becomes highly reactive and the full number of bonds is soon restored by a reaction involving bonding with another atom or molecule. In a pure sample of an element, the valencies are satisfied by bonding between atoms of the same kind. The only elements that exist in their normal states as single, unattached atoms are the noble gases (group 0, Periodic Table), which are elements of zero valency.

This electronic picture of valency is not necessary for visualizing the bonding in a compound of known formula and structure. The formula can be written as a diagram in which the atoms are circles, a line attached to one circle is a valency and a line attached at both ends is a chemical bond.

Thus water, $H_2O$, is composed of hydrogen, —Ⓗ, and oxygen, Ⓞ⟨ .

The formula can be drawn, so that there are no unsatisfied valencies, in the following manner:

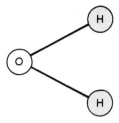

This picture tells us which atom is attached to which.
Slightly more complicated is rust, ferric oxide, $Fe_2O_3$.

Ferric iron, Ⓕ<, and oxygen, Ⓞ<, are combined in the following manner:

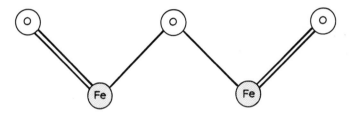

Once familiar with the system, we can dispense with the circles and abbreviate the picture. An example of variable valency in one compound is nitrous oxide, $N_2O$ ('laughing gas'), with nitrogen valencies of 3 and 5:

$$N \equiv N = O$$

It is occasionally possible to work out a formula from scant information. Asked to find a formula for calcium chloride, we assume that it contains only calcium, Ca<, and chlorine, —Cl. Thus the formula must be $CaCl_2$ and cannot be CaCl or $CaCl_3$.

## Electronegativity

We have seen that bonding is a consequence of the attraction that the positive nucleus of an atom has for the negative electrons of other atoms that come within its 'sphere of influence'. This attraction is stronger in some atoms than others. It depends upon the number of positive charges (protons) in the nucleus, the distance from the nucleus to the extremities of the atom and the number of non-valency electrons in between. These electrons provide a negative 'screen' round the nucleus, reducing its power of attraction for valency electrons. We say that those elements with a strong attraction for valency electrons are highly *electronegative*. The American chemist Pauling has calculated numerical values for electronegativity and placed the elements in a series, in which those with the highest values are the most electronegative.

*Part of Pauling's Scale of Electronegativity*

| F | 4·0 | S | 2·5 | B | 2·0 | Ca | 1·0 |
|---|-----|---|-----|----|-----|----|-----|
| O | 3·5 | C | 2·5 | Si | 1·8 | Na | 0·9 |
| Cl | 3·0 | I | 2·4 | Sn | 1·7 | Ba | 0·9 |
| N | 3·0 | P | 2·1 | Al | 1·5 | K | 0·8 |
| Br | 2·8 | H | 2·1 | Mg | 1·2 | | |

We should note that the last seven elements are all metals and the first twelve all non-metals. Also that oxygen is particularly greedy for electrons and nitrogen only a little less so. Carbon and hydrogen are moderately electronegative, but do not differ much in their attractive power for electrons. Electronegativity determines much of the behaviour of elements and their compounds.

## Chemical reactions and equations

Let us examine bonding in action and see how the process – chemical reaction – is written down. If the two elements iron and sulphur are heated together a dark powder is produced, ferrous sulphide. Sulphur has a valency of two; iron, in this instance, is in its lower (ferrous) valency state, also two.

This pictorial representation can be replaced by symbols only, to give the neater description:

$$Fe + S \longrightarrow FeS$$

This is a *chemical equation*. It states how many atoms (or molecules) of each substance are required and how many are produced. Since it is an equation, the two sides equal one another; they contain the same number of atoms of each element, but differently combined. This is because, in a chemical reaction, matter is never destroyed; it is simply converted into something else. Thus burnt carbon 'disappears' by conversion into the transparent gas, carbon dioxide:

$$C + O_2 \longrightarrow CO_2$$

In chemical equations, elementary gases are always written as molecules rather than atoms; hence $O_2$ and not $2O$ (two atoms).

Some reactions involve several molecules of each substance:

$$4Fe + 3O_2 + 2H_2O \longrightarrow 2Fe_2O_3 \cdot H_2O$$

This equation contains 4 atoms of iron, 8 of oxygen and 4 of hydrogen on each side, i.e. it 'balances'. It denotes that two molecules of ferric oxide are produced from four atoms of iron and three molecules of oxygen and that each molecule of ferric oxide has one molecule of water associated with it in the solid crystal.

Reactions also occur between compounds. Common salt is made by mixing sodium hydroxide and hydrochloric acid.

$$Na{\mid}OH \;+\; H{\mid}Cl \longrightarrow NaCl + H_2O$$

sodium    hydrochloric
hydroxide     acid

Note that in this and many other reactions between compounds, the reaction simply involves exchanging halves of the molecules. This is known as *double decomposition*. Again

$$Ca{\mid}CO_3 \;+\; 2H{\mid}Cl \longrightarrow Ca{\mid}Cl_2 \;+\; H_2{\mid}CO_3$$

calcium                  carbonic
carbonate               acid

Carbonic acid decomposes at normal temperatures:

$$H_2CO_3 \longrightarrow CO_2 + H_2O$$

Note that the half exchanged can consist of a group of atoms. Groups that frequently occur are:

| Name | Formula | Valency |
|------|---------|---------|
| Carbonate | $=CO_3$ | 2 |
| Bicarbonate | $-HCO_3$ | 1 |
| Sulphate | $=SO_4$ | 2 |
| Nitrate | $-NO_3$ | 1 |
| Hydroxide | $-OH$ | 1 |

The group valency is the valency left over after the group atoms are bonded together, e.g.

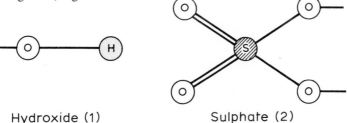

Hydroxide (1)                      Sulphate (2)

Since the groups are found only in compounds, the valencies are never left unsatisifed.

# Two

# Inorganic chemistry

Inorganic Chemistry is the study of all the elements except carbon, though that element itself and some few simple carbon compounds are included under the heading. We have space here only to define some important ideas in inorganic chemistry and to have a brief look at the compounds of the four elements that we will meet most frequently in Part II: carbon, hydrogen, oxygen and nitrogen.

## Acids, bases and salts

All compounds are either acidic, basic or neutral. The nature of a substance soluble in water is easily determined. *Pure* water is neutral and has no effect on litmus paper, be it red or blue. A solution of an acid, however, turns blue litmus red, while a basic solution turns red litmus blue. Should we place two inert electrodes in the solution (called the *electrolyte*) and allow a small direct current to flow through it, then hydrogen bubbles will appear at the negative electrode (the *cathode*) if the solution is acid; if it is basic, then oxygen bubbles will appear at the positive electrode (the *anode*). This process is known as *electrolysis*.

It is accounted for by the existence of electrically charged atoms or groups, known as *ions*. If a chemical bond is formed between two atoms of elements at the *opposite extremes* of the electronegativity series, then the following can occur. The more electronegative element exerts a very strong attraction on the valency electron from the weakly electronegative element and that element has little counter-attraction to retain it. The electron is therefore completely absorbed into the 'sphere of influence' of the electronegative atom and becomes part of it. The atom is now negatively charged, having more electrons than protons, and being charged, is now called an ion (an *anion,* because it discharges at the anode). The atom which has lost an electron becomes a positively charged ion (a *cation,* because it discharges at the cathode). Between the two ions of opposite charge exists a strong electrostatic attraction and this bond between them is called an *ionic bond.*

An ionic bond may also be formed between two groups of atoms, or between an atom and a group. It is not always easy to predict such an ionic bond from the molecular formula, since the electronegativity of the main atom in a group is modified by the other atoms covalently bonded to it.

If some of the solid ionic substance is put into water, then the high dielectric constant, or insulating property, of that liquid permits the ions to *dissociate* from one another and exist separately in the solution. Once the electric current is switched on, the ions are attracted by and move towards the electrode of opposite charge. They reach it and are discharged.

We can see why the electrolysis products are what they are, when we known that all acidic solutions contain hydrogen ions, $H^+$, and all basic solutions contain hydroxide ions, $OH^-$. In the hydroxide ion, the electronegative oxygen is responsible for the charge. Thus for acids, at the cathode (negative), cations discharge by gaining electrons:

$$2H^+ + 2\,\text{electrons} \longrightarrow 2H \longrightarrow H_2 \uparrow$$

and for bases, at the anode (positive), anions discharge by losing electrons:

$$4OH^- - 4\,\text{electrons} \longrightarrow 4OH \longrightarrow 2H_2O + 2O \longrightarrow O_2 \uparrow$$

Hydrochloric acid is clearly capable of producing hydrogen ions by dissociation:

$$HCl \rightleftharpoons H^+ + Cl^-$$

The two-way double arrow is written to indicate that while ions are forming, some are recombining to form molecules. The preponderance of ions over molecules is indicated by the longer arrow in that direction. Sulphur trioxide, $SO_3$, is an acidic gas, though it contains no hydrogen. In water it is converted to sulphuric acid:

$$SO_3 + H_2O \longrightarrow H_2SO_4 \rightleftharpoons 2H^+ + SO_4^{2-}$$

The two negative charges on the sulphate ion are obtained via the two bonds between separate oxygen atoms and the two hydrogen atoms.

Again, sodium hydroxide (caustic soda), NaOH, gives hydroxide ions by

$$NaOH \rightleftharpoons Na^+ + OH^-$$

but ammonia gas, $NH_3$, is basic because in solution

$$NH_3 + H_2O \longrightarrow NH_4OH \rightleftharpoons -NH_4^+ + OH^-$$
$$\text{ammonium}$$
$$\text{hydroxide}$$

The reverse direction of the long arrow indicates little dissociation.

Water itself is very weakly dissociated

$$H_2O \rightleftharpoons H^+ + OH^-$$

but is neutral because the numbers of hydrogen and hydroxide ions exactly balance.

## Acids

The best known strong acids are the three mineral acids:

| HCl | $HNO_3$ | $H_2SO_4$ |
|-----|---------|-----------|
| hydrochloric acid | nitric acid | sulphuric acid |

The first two are the strongest, strength being the degree of dissociation to $H^+$ ions in solution. They are highly corrosive in concentrated solution and sulphuric acid in particular has the ability to dilute itself, by extracting water from the structures of other chemicals. Orthophosphoric acid, $H_3PO_4$, is a moderately strong acid. Chromic acid, $H_2CrO_4$, exists only in aqueous solutions.

Weak acids are usually organic chemicals, e.g. acetic acid (vinegar) and citric acid (fruit juices). Exceptions are nitrous ($HNO_2$) and sulphurous ($H_2SO_3$) acids, which are stable only in solution, where they are slightly dissociated. Solutions of carbon dioxide behave as weak acids, because they have extremely low carbonic acid contents:

$$CO_2 + H_2O \rightleftharpoons H_2CO_3$$

## Bases

The strong bases are the *alkalis*, such as sodium and potassium hydroxides (NaOH and KOH). Strength is the degree of dissociation to $OH^-$ ions in solution. In concentrated solutions these are also corrosive. Weak bases include ammonia (above) and its derivatives in Organic Chemistry (Chapter 4).

## Salts

When an acid and a base are mixed, chemical reaction occurs and a salt and water are formed:

$$2KOH + H_2SO_4 \longrightarrow K_2SO_4 + H_2O$$
$$\text{potassium}$$
$$\text{sulphate}$$

Salts take their names from the parent acids:

| Acid | Salt |
|------|------|
| Nitric acid | Nitrate |
| Nitrous acid | Nitrite |
| Sulphuric acid | Sulphate or bisulphate |
| Sulphurous acid | Sulphite or bisulphite |
| Hydrochloric acid | Chloride |
| Carbonic acid | Carbonate or bicarbonate |
| Hydrogen sulphide | Sulphide |
| Orthophosphoric acid | Orthophosphate |
| Chromic acid | Chromate |

A salt with 'bi' in the name contains an atom of hydrogen from the original acid, e.g. sodium bisulphite, $NaHSO_3$. However, phosphoric acid with three hydrogen atoms forms primary ($NaH_2PO_4$), secondary ($Na_2HPO_4$) and tertiary ($Na_3PO_4$) phosphates.

Salts dissociate in water to produce ions. Potassium sulphate is neutral

$$K_2SO_4 \longrightarrow 2K^+ + SO_4^{2-}$$

Thus if there is no excess of $H_2SO_4$ or KOH molecules after the formation of $K_2SO_4$, the acid and alkali *neutralize* one another. The formation of the salt can be considered as the mutual neutralization of the $H^+$ and $OH^-$ ions to form water:

$$\underset{\text{(from } H_2SO_4)}{H^+} + \underset{\text{(from KOH)}}{OH^-} \rightleftharpoons H_2O$$

Other salts are not necessarily neutral. Ammonium chloride, $NH_4Cl$, has an acid solution:

$$NH_4Cl \rightleftharpoons NH_4^+ + Cl^- \text{ and } H_2O \rightleftharpoons H^+ + OH^-$$

but $OH^-$ ions from water are removed by combination with $NH_4^+$ to form the weakly dissociated ammonium hydroxide. This leaves $H^+$ and $Cl^-$ ions in solution, i.e. a dilute solution of fully dissociated hydrochloric acid. The salt is formed from a weak base and a strong acid: in solution it is the $H^+$ ions from the latter which are in excess over $OH^-$ and give the salt its acid character.

Similarly, sodium bisulphite, $NaHSO_3$, although it contains hydrogen (which might be thought to be a source of $H^+$ ions), has a basic solution:

$$NaHSO_3 \rightleftharpoons Na^+ + HSO_3^- \text{ and } H_2O \rightleftharpoons H^+ + OH^-$$

but $H^+$ ions from water are removed by combination with bisulphite ions to form the weakly dissociated sulphurous acid. This leaves $Na^+$ and $OH^-$ ions in solution, i.e. a dilute solution of fully dissociated sodium hydroxide. Sodium bisulphite is the salt of a strong base and a weak acid.

These examples emphasize that acidity is the ability to produce $H^+$ ions in solution and depends on no other property. Basicity is associated with $OH^-$ ions. The more $H^+$ ions produced, the stronger the acid. Acidic or basic strength is measured on a scale called the *pH scale*. A pH of 7 is exactly neutral, 1 is strongly acid and 13 strongly alkaline.

## Titration

By the use of *indicators* – compounds which change colour as their environment changes from an acidic to a basic one (or vice versa) – the precise moment of neutralization can be determined. For example, a measured quantity of acid solution of unknown concentration is placed in a flask with an indicator (e.g. litmus). A solution of base of known concentration is run into the flask until, with the addition of *one excess drop* of base, the colour changes. The volume of base run in has been measured and it is now possible to calculate the concentration of the acid solution. The method of calculation can be found in any elementary chemistry book and depends on knowledge of the *equivalent weights* of acids and bases. Titrations are used in paint chemistry to determine the Acid Value of a resin (Chapter 12) and it will be seen from the definition of Acid Value that no knowledge of equivalent weights is required for this determination; it is sufficient to know the concentration of the alkali solution.

**Other acidic and basic compounds**

A large number of materials – particularly oxides – are not obviously acidic or basic from their formulae, but yet behave in an acidic or basic manner. The oxides of non-metals are invariably neutral or *acidic*. The acidic oxides form acids in water and neutralize alkalis, e.g.:

$$2NaOH + B_2O_3 \longrightarrow 2NaBO_2 + H_2O$$
$$\text{boric} \qquad\qquad \text{sodium}$$
$$\text{oxide} \qquad\qquad \text{borate}$$

Most metallic oxides are *basic* and neutralize acids, e.g.

$$Fe_2O_3 + 6HCl \longrightarrow 2FeCl_3 + 3H_2O$$
$$\text{ferric}$$
$$\text{chloride}$$

but a number of metallic oxides can neutralize both acids and bases and, because of this double character, are called *amphoteric oxides*, e.g.

$$Al_2O_3 + 6HCl \longrightarrow 2AlCl_3 + 3H_2O$$
$$\text{aluminium} \qquad\qquad \text{aluminium}$$
$$\text{oxide} \qquad\qquad\quad \text{chloride}$$

$$2NaOH + Al_2O_3 \longrightarrow 2NaAlO_2 + H_2O$$
$$\text{sodium}$$
$$\text{aluminate}$$

## Polarity and the hydrogen bond

If a bond exists between two atoms of similar electronegativity, e.g. C and H, the electrons in the bond are fairly evenly shared and reasonably centrally situated between the two atomic nuclei. Such a bond is said to be *non-polar*. If, however, there is a marked difference in electronegativity, e.g. C and O or H and O, the electrons are displaced towards the more electronegative atom. Ionization may not occur if the displacement is insufficient, but the bond will not be electrically symmetrical, e.g. $\overset{+}{C} :\overset{-}{O}$, where : are the electrons in the bond. Such an arrangement of atoms is said to be *polar* and gives the group a slight electrical charge as shown.

In a polar $\overset{-}{O}\!-\!\overset{+}{H}$ bond, the hydrogen atom, which consists solely of one proton and one electron, has practically become a hydrogen ion, i.e. a proton. Should another electronegative atom in a polar group come near this proton, there is an attraction between the negative and positive charges sufficient to form a weak chemical bond between the two groups. In water, for example:

$$\overset{+}{H}\!-\!\overset{-}{O}$$
$$\overset{+}{H}\!:\!-\!\overset{-}{O}\!-\!\overset{+}{H}$$
$$\overset{|}{H}^{+}$$

The *hydrogen bond*, as it is called, is shown by a broken line. The effect of this weak bonding between molecules is to raise the boiling point, since heat energy is required to break the bond before the molecules can be separated. Thus $H_2O$ boils at 100 °C, while $H_2S$ (sulphur is immediately below oxygen in the Periodic Table), though a heavier molecule, boils at −61 °C. Electronegativities are O 3·5, S 2·5 and H 2·1. Hydrogen bonding affects solvency in paints (Chapter 9).

## Oxygen and hydrogen

### Oxidation and reduction

The chemistry of these elements is closely associated with a very important type of chemical reaction: *oxidation and reduction.* An atom which has been oxidized is one that, in the course of a chemical reaction, loses control – totally or partially – over one or more of its valency electrons, in favour of a more electronegative atom. Reduction *occurs simultaneously,* since the reduced atom gains control over an electron from a less electronegative atom.

This definition may become clearer after some examples, but in the process, simpler practical definitions can be made. Thus, *oxidation is the addition of oxygen to an element or compound,* e.g.

$$2Mg + O_2 \xrightarrow{\text{burns}} 2MgO$$
$$\text{magnesium}$$
$$\text{oxide}$$

$$\text{and} \quad 2SO_2 + O_2 \xrightarrow[\substack{Pt \\ \text{catalyst}}]{\text{heat}} 2SO_3$$

*Reduction is* the opposite process: *the removal of oxygen from a compound.* Hydrogen is often used to do this, e.g.

$$CuO + H_2 \xrightarrow{\text{heat}} Cu + H_2O$$
$$\text{copper}$$
$$\text{oxide}$$

In these reactions, oxygen is the *oxidizing agent* and hydrogen the *reducing agent*. In the last example, while the CuO is reduced, the hydrogen is oxidized to water and in the other examples, oxygen is reduced to an oxide. The processes of oxidation and reduction always occur together.

The oxidizing agent need not be oxygen, e.g.

$$C + 2H_2SO_4 \longrightarrow CO_2 + 2H_2O + 2SO_2$$

Carbon is oxidized to $CO_2$ and $H_2SO_4$ is reduced to $SO_2$. Similarly, hydrogen need not be the reducing agent. Carbon is in the above example and again in

$$PbO + C \xrightarrow{\text{heat}} Pb + CO$$
$$\text{lead} \qquad\qquad\qquad \text{carbon}$$
$$\text{monoxide} \qquad\qquad\quad \text{monoxide}$$
$$\text{(litharge)}$$

While *oxidation is* the addition of oxygen, it is *also the removal of hydrogen,* e.g.

$$Cl_2 + H_2S \longrightarrow 2HCl + S$$

$H_2S$ is oxidized to sulphur by chlorine (the oxidizing agent), which is itself *reduced* to HCl *by the addition of hydrogen.*

Finally, if an element of variable valency is caused to increase its valency, it has been oxidized, but if its valency is reduced, the element has been reduced, e.g.

$$2FeCl_2 + Cl_2 \longrightarrow 2FeCl_3 \qquad \textit{oxidation}$$
$$\text{Fe} = 2 \qquad\qquad\quad \text{Fe} = 3$$
$$\text{ferrous chloride} \qquad \text{ferric chloride}$$

$$2FeCl_3 + H_2 \longrightarrow 2FeCl_2 + 2HCl \quad \textit{reduction}$$
$$\text{Fe} = 3 \qquad\qquad\quad \text{Fe} = 2$$

The original definition is most easily understood if applied to the last example. Consider the ionizations

$$FeCl_2 \rightleftharpoons Fe^{2+} + 2Cl^-$$
$$FeCl_3 \rightleftharpoons Fe^{3+} + 3Cl^-$$

Oxidation changes $Fe^{2+}$ to $Fe^{3+}$: loss of one electron. Reduction returns that electron. The other examples all fit the definition, though where ionization does not occur, electronegativies must be used to see which element has lost and which has gained some control over a valency electron.

*OXYGEN* is made in the laboratory by heating potassium chlorate.

$$2KClO_3 \longrightarrow 2KCl + 3O_2$$

10 per cent of manganese dioxide, $MnO_2$, is added to speed up the reaction. It is unchanged by the heating and can be recovered afterwards. This ability

to accelerate the reaction without being used up in the process is called *catalysis* and the substance with this ability is a *catalyst*. Industrially, oxygen is separated from nitrogen and other atmospheric gases by cooling to the liquid state. Its boiling point, $-182 \cdot 5\,°C$, differs from those of the other gases.

*HYDROGEN* is made by the reaction between acids and metals, e.g.

$$Zn + H_2SO_4 \longrightarrow ZnSO_4 + H_2$$
$$\text{(dilute)} \qquad \text{zinc}$$
$$\text{sulphate}$$

### Hydrogen peroxide

Oxygen and hydrogen are found in countless chemical compounds. One of the most interesting is their joint compound, *hydrogen peroxide*, $H_2O_2$. It is made by several methods one of which is

$$BaO_2 + H_2SO_4 \xrightarrow{\;0\,°C\;} BaSO_4 \downarrow + H_2O_2$$
$$\text{barium} \quad \text{(dilute)} \qquad \text{barium}$$
$$\text{peroxide} \qquad\qquad \text{sulphate}$$

The barium sulphate is insoluble in water and settles out.

The valencies indicate that the atoms must be connected thus:

$$H\!-\!O\!-\!O\!-\!H$$

The compound is not very stable and is decomposed by heat and light. When a compound decomposes, chemical bonds are broken. A covalent bond may break in two ways: either the two electrons in the bond will be shared between the atoms joined by the bond (homolytic dissociation), or they will be retained by the more electronegative of those atoms (heterolytic dissociation).

(i) homolytic $A:B \longrightarrow A. + .B$

(ii) heterolytic $A:B \longrightarrow A^+ + :B^-$

Heterolytic dissociation produces ions. Homolytic dissociation – which is the more likely method if the two fragments are equally electronegative – produces neutral atoms or groups, each with an unsatisfied valency or unpaired valency electron. This electron is denoted by a dot. Such groups are known as *free radicals*. Most free radicals are highly reactive, satisfying their valencies by forming new chemical bonds at the earliest opportunity.

$H\!-\!O\!-\!O\!-\!H$ is a very symmetrical formula, so that the usual method of decomposition is homolytic:

$$H_2O_2 \longrightarrow 2HO\cdot$$

The final decomposition products are:

$$2H_2O_2 \longrightarrow 2H_2O + O_2$$

unless some other chemical is present with which the free radicals may react. Since the oxygen in the free radical requires another electron to form a bond, and oxygen is highly electronegative, $H_2O_2$ is a powerful oxidizing agent, e.g.

$$PbS + 4H_2O_2 \longrightarrow PbSO_4 + 4H_2O$$
$$\text{lead} \qquad\qquad\qquad \text{lead}$$
$$\text{sulphide (black)} \qquad \text{sulphate (white)}$$

This reaction is used for 'cleaning' paintings, where lead pigments have discoloured.

## Carbon

Although the element is found in many impure forms, such as soot, lamp-black, charcoal, coal and coke, the pure substance exists in two very different forms: diamond and graphite (the substance used with clay to form pencil 'leads').

The outstanding property of the element carbon is its ability to bond *with itself* to form chains of atoms. Other elements joined in this way give unstable compounds after a chain of 3 or 4 atoms, but carbon appears to have no limit. The wide variety of atoms and groups that may be attached to the valencies not forming the chain, leads to an infinite variety of carbon compounds, whose study is the subject 'Organic Chemistry'.

The inorganic compounds of carbon include the oxides CO and $CO_2$, carbonic acid and its salts and the carbides.

In a restricted supply of air

$$2C + O_2 \xrightarrow{\;800\,°C\;} 2CO$$

*Carbon monoxide* is a neutral gas, highly dangerous because it is poisonous and has no smell. It is used as a fuel in coal gas.

$$2CO + O_2 \longrightarrow 2CO_2$$

*Carbon dioxide* is more conveniently made from dilute acid and a carbonate

$$CaCO_3 + 2HCl \longrightarrow CaCl_2 + CO_2 + H_2O$$

The gas is easily liquefied by compression and becomes solid at $-79\,°C$. The 'dry ice' formed is used for cooling. The compressed gas is used in fire extinguishers, since nothing will burn in it. Water dissolves almost its own volume of $CO_2$ at normal temperatures and pressure (*carbonic acid*). With alkali

$$2NaOH + CO_2 \longrightarrow Na_2CO_3 + H_2O$$
$$\text{sodium carbonate}$$
$$\text{(washing soda)}$$

$$\text{and } Na_2CO_3 + H_2O + CO_2 \longrightarrow 2NaHCO_3$$
$$\text{sodium bicarbonate}$$
$$\text{(baking powder)}$$

These salts dissolve in water, together with potassium and ammonium carbonates. *Carbonates* of metals outside Group IA in the Periodic Table are insoluble:

$$Pb(NO_3)_2 + Na_2CO_3 \longrightarrow PbCO_3 \downarrow + 2NaNO_3$$

lead · · · · · · · · · · · · · · · · · · · · lead · · · · · sodium
nitrate · · · · · · · · · · · · · · · · · carbonate · nitrate

All the other salts in the equation are soluble. Alkali metal carbonates are stable, but others decompose on heating, e.g.

$$PbCO_3 \longrightarrow PbO + CO_2$$

Sodium and potassium *bicarbonates* are less soluble than the carbonates. Magnesium, calcium, strontium and barium bicarbonates exist only in solution. Attempts to remove the water destroy the compounds

$$Ca(HCO_3)_2 \xrightarrow{100\,°C} CaCO_3 \downarrow + H_2O + CO_2$$

calcium bicarbonate · · · · · · · ("fur" in a kettle)
(temporary hardness
in water)

The alkali bicarbonates are also decomposed by heat.
    Some metal oxides form *carbides* when heated with carbon (coke)

$$CaO + 3C \xrightarrow{2000\,°C} CaC_2 + CO$$

calcium oxide · · · · · · · · · · calcium
(lime) · · · · · · · · · · · · · · · · · carbide

This compound produces acetylene gas by reaction with water

$$CaC_2 + 2H_2O \longrightarrow Ca(OH)_2 + C_2H_2$$

calcium · · · · acetylene
hydroxide
(slaked lime)

## Nitrogen

This is an element essential to plant growth. It can be obtained from the air, either by removing the oxygen by burning it to form a solid oxide, or by liquefying the air (see *Oxygen*). In the laboratory it is made thus:

$$NH_4NO_2 \xrightarrow{heat} N_2 + 2H_2O$$

ammonium
nitrite

It is a colourless, odourless, neutral gas. It is very unreactive, but forms *nitrides* when strongly heated with some metals, e.g.

$$3Mg + N_2 \longrightarrow Mg_3N_2$$
$$\text{magnesium}$$
$$\text{nitride}$$

and can be made to react with hydrogen to form the basic gas, *ammonia.*

$$N_2 + 3H_2 \xrightarrow[\substack{\text{very high} \\ \text{pressure}}]{500\,°C} 2NH_3$$

Salts of ammonium hydroxide, $NH_4OH$, all contain the ammonium ion, $NH_4^+$. On heating with a base, they give ammonia

$$2NH_4Cl + Ca(OH)_2 \longrightarrow 2NH_3 + CaCl_2 + 2H_2O$$

The chief *oxides of nitrogen* are made as follows:

$$NH_4NO_3 \xrightarrow{\text{heat}} N_2O + 2H_2O \quad \text{a neutral oxide}$$
$$\text{ammonium} \qquad\qquad \text{nitrous}$$
$$\text{nitrate} \qquad\qquad\quad \text{oxide}$$
$$\text{(laughing gas)}$$

$$3Cu + 8HNO_3 \longrightarrow 3Cu(NO_3)_2 + 2NO + 4H_2O$$
$$(50\%) \qquad\qquad\qquad\qquad\quad \text{nitric}$$
$$\text{oxide}$$

Nitric oxide is a neutral, colourless gas, but mere contact with oxygen oxidizes it to the acidic, brown gas, nitrogen peroxide. At normal temperatures, this is a mixture of 'single' and 'double' molecules:

$$2NO + O_2 \longrightarrow N_2O_4 \underset{-10\,°C}{\overset{140\,°C}{\rightleftharpoons}} 2NO_2$$

$$2NO_2 \text{ (or } N_2O_4) + H_2O \longrightarrow HNO_3 + HNO_2$$

*Nitrous acid* is stable only at lower temperatures and decomposes thus:

$$3HNO_2 \longrightarrow HNO_3 + H_2O + 2NO$$
$$\downarrow +O_2$$
$$2NO_2 \text{ etc.}$$

The salts (*nitrites*) are more stable than the acid.

*Nitric acid* is made industrially via NO

$$\text{(a)} \quad 4NH_3 + 5O_2 \xrightarrow{\text{red hot Pt}} 4NO + 6H_2O \text{ (Pt is a catalyst)}$$

$$\text{or} \quad \text{(b)} \quad N_2 + O_2 \xrightarrow[\text{high temp.}]{\text{electric arc}} 2NO$$

Dilute nitric acid shows normal acidic properties, but the concentrated acid is also an oxidizing agent. Note the action on copper (above) when hydrogen is *not* produced (see normal acid-metal reaction under *Hydrogen*).

*Nitrates* differ in the way they decompose on heating. $NH_4NO_3$ is shown above. Alkali nitrates give oxygen:

$$2KNO_3 \longrightarrow 2KNO_2 + O_2$$

potassium nitrate (saltpetre) → potassium nitrite

other nitrates lose oxygen *and* nitrogen peroxide:

$$2Pb(NO_3)_2 \longrightarrow 2PbO + 4NO_2 + O_2$$

# Three

# Organic chemistry: paraffins to oils

Organic Chemistry is the study of carbon compounds. The principal element associated with carbon in these compounds is hydrogen. The majority of the compounds of interest here contain these elements and only two others: oxygen and nitrogen.

This chapter is concerned solely with *aliphatic* compounds – those deriving from fatty materials. The simplest of these compounds contain carbon and hydrogen only.

## Aliphatic hydrocarbons

### The paraffins

The valency of carbon is four, that of hydrogen, one. The simplest possible

compound is, therefore: $H-\overset{\displaystyle H}{\underset{\displaystyle H}{C}}-H$ . With a chain of two carbon atoms we

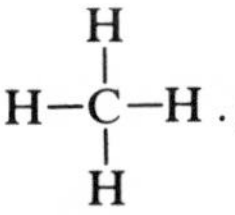

methane

have $H-\overset{\displaystyle H}{\underset{\displaystyle H}{C}}-\overset{\displaystyle H}{\underset{\displaystyle H}{C}}-H$ , three $H-\overset{\displaystyle H}{\underset{\displaystyle H}{C}}-\overset{\displaystyle H}{\underset{\displaystyle H}{C}}-\overset{\displaystyle H}{\underset{\displaystyle H}{C}}-H$ and so on. Note that

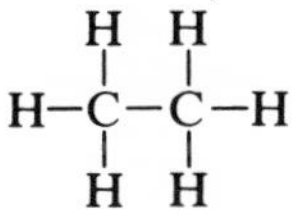

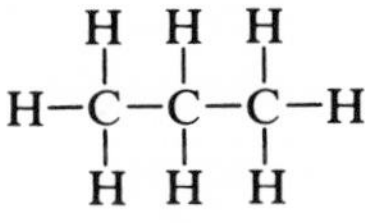

ethane                       propane

in each compound the number of H atoms is *twice* the number of C atoms *plus two,* so the general formula for the paraffins is $C_nH_{2n+2}$ (where $n$ is any whole number).

It is important at once to get a true picture of what these molecules look like in three dimensions. We must realize that in these compounds the carbon valencies have fixed directions and the bonds are equi-distant in space, pointing from the centre to the corners of a tetrahedron (see Fig. 4).

Thus in propane (Fig. 5), the carbon atoms are not connected together in a straight line as in the two-dimensional formula.

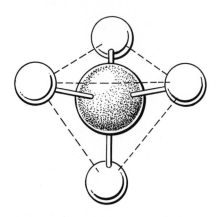

**Fig. 4** A methane molecule.

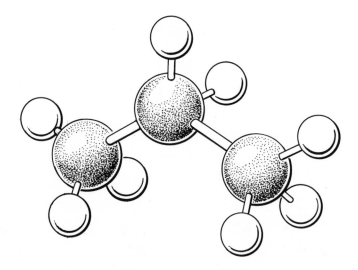

**Fig. 5** A propane molecule.

Furthermore, the other attached groups or atoms can rotate about any single bond (Fig. 6). A model molecule can be made from wooden balls (atoms) containing sockets, into which metal rods (chemical bonds) may be fitted. Since the rods fit loosely rotation is easily demonstrated.

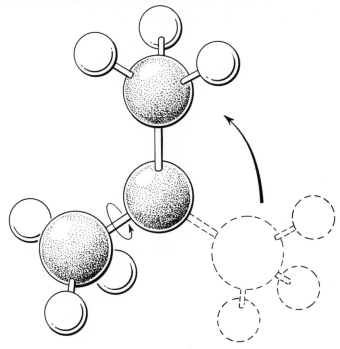

**Fig. 6** Rotation about a single bond. The hydrogen atoms on the central carbon atom are omitted to simplify the figure.

Thus a long hydrocarbon-chain is able to coil by rotation around the single bonds (Fig. 7).

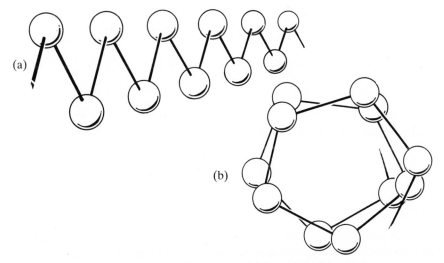

**Fig. 7** A chain of carbon atoms: (a) extended, (b) coiled.

Coming back to the series of paraffins, a fourth carbon atom could be attached to propane at an end carbon or at the centre carbon. Thus two $C_4$ paraffins are possible:

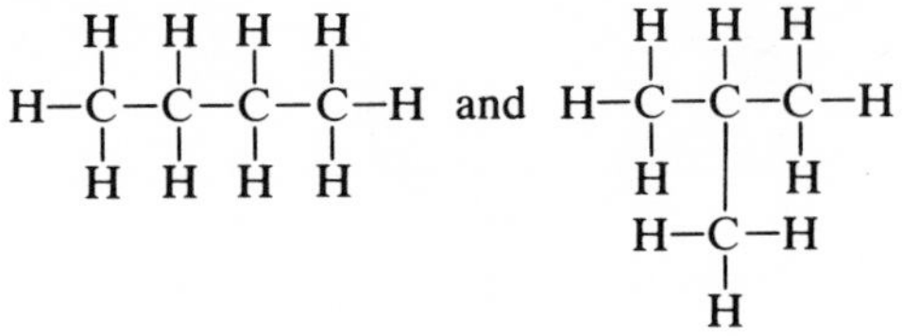

$$
\begin{array}{ccccc}
 & H & H & H & H \\
 & | & | & | & | \\
H- & C- & C- & C- & C-H \\
 & | & | & | & | \\
 & H & H & H & H
\end{array}
\quad \text{and} \quad
\begin{array}{cccc}
 & H & H & H \\
 & | & | & | \\
H- & C- & C- & C-H \\
 & | & | & | \\
 & H & & H \\
 & & H-C-H & \\
 & & | & \\
 & & H &
\end{array}
$$

*n* (normal)-butane           *iso*-butane

*n*-Butane is a *straight chain* compound, *iso*-butane contains a *branched chain*. Both have the overall formula, $C_4H_{10}$, but the atoms are differently arranged. This is enough to give the compounds some differences in properties, e.g. boiling points: *n*-butane $-0.5\,°C$, *iso*-butane $-11.7\,°C$. Compounds with the same overall formula, but a different structure, are called *isomers*. The straight chain paraffin isomer is always called the *normal* form.

In Organic Chemistry there are so many isomers that overall formulae are useless and are not used. Structural formulae are used but, for ease of printing, every bond is not shown. The butanes, for example, are abbreviated to:

$$CH_3 \cdot CH_2 \cdot CH_2 \cdot CH_3$$
or
$$CH_3 \cdot (CH_2)_2 \cdot CH_3$$

*n*-butane

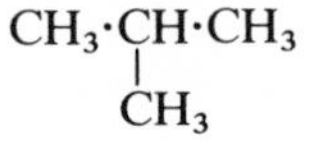

$$CH_3 \cdot CH \cdot CH_3$$
$$|$$
$$CH_3$$
or
$$CH_3 \cdot CH(CH_3) \cdot CH_3$$

*iso*-butane

With five carbons in the molecule there are three isomers

$$CH_3 \cdot CH_2 \cdot CH_2 \cdot CH_2 \cdot CH_3$$

*n*-pentane

$$CH_3 \cdot CH_2 \cdot CH(CH_3) \cdot CH_3$$

*iso*-pentane

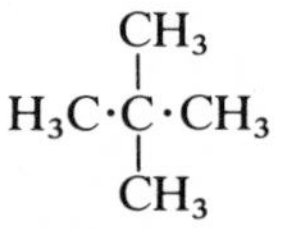

$$
\begin{array}{c}
CH_3 \\
| \\
H_3C \cdot C \cdot CH_3 \\
| \\
CH_3
\end{array}
$$

*neo*-pentane

and formation of a stable ring is possible:

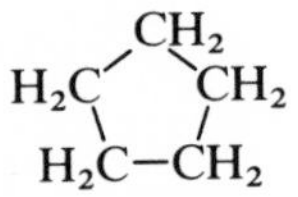

$$
\begin{array}{c}
CH_2 \\
H_2C \quad CH_2 \\
H_2C-CH_2
\end{array}
$$

*cyclo*-pentane

There is no limit to the number of carbon atoms. More complicated structures take their names from the longest straight chain of carbon atoms, which is numbered, and the side chains are described as being attached to such-and-such number carbon atoms, e.g.

$$
\begin{array}{cccccc}
1 & 2 & 3 & 4 & 5 & 6 \\
CH_3-CH-CH_2-CH-CH_2-CH_3 \\
\quad\quad | \quad\quad\quad | \\
\quad\quad CH_3 \quad\quad CH-CH_3 \\
\quad\quad\quad\quad\quad\quad | \\
\quad\quad\quad\quad\quad\quad CH_3
\end{array}
$$

is 2-methyl 4-*iso*-propyl hexane, even though, with 10 carbon atoms, it is a complex decane. Note that it is 2,4-substituted, not 3,5-, since the order of numbering is chosen to give the lowest numbers. Also note the -yl ending for the attached groups: methyl from methane and *iso*-propyl from *iso*-propane. These groups are *alkyl* groups, since they are derived from alkanes (paraffins).

The physical state of the paraffin alters as the number of carbon atoms increases. Heavier molecules require more energy to escape from their neighbours:

| Paraffin | Formula | Boiling point °C | Melting point °C |
|---|---|---|---|
| Methane | $CH_4$ | − 161 | |
| Butane | $CH_3 \cdot (CH_2)_2 \cdot CH_3$ | − 0·5 | |
| Pentane | $CH_3 \cdot (CH_2)_3 \cdot CH_3$ | 36 | |
| *iso*-Pentane | $CH_3 \cdot CH_2 \cdot CH(CH_3)_2$ | 28 | |
| *neo*-Pentane | $C \cdot (CH_3)_4$ | 9 | |
| Pentadecane | $CH_3 \cdot (CH_2)_{13} \cdot CH_3$ | 273 | 10 |
| Octadecane | $CH_3 \cdot (CH_2)_{16} \cdot CH_3$ | 318 | 28 |

Chief natural sources of paraffins are *natural gas* (chiefly methane) and *petroleum*. The mixture in petroleum is so complex that the paraffins are not completely separated from one another, but are collected in groups according to *boiling range,* by distillation. The following 'fractions' are collected:

| Boiling range | Fraction | C content | Use |
|---|---|---|---|
| 20–100 °C | light petrol | $C_5–C_7$ | solvent (see Chapter 9) |
| 70–90 °C | benzine | $C_6–C_7$ | dry cleaning |
| 80–120 °C | ligroin | $C_6–C_8$ | solvent (see Chapter 9) |
| 70–200 °C | petrol | $C_6–C_{11}$ | motor fuel |
| 200–300 °C | paraffin oil (kerosene) | $C_{12}–C_{16}$ | lighting |
| Above 300 °C | gas oil (heavy oil) | $C_{13}–C_{18}$ | fuel oil |
| | lubricating oil | $C_{16}–C_{20}$ | lubricant |
| | greases, vaseline | $C_{18}–C_{22}$ | pharmaceutical |
| | paraffin wax | $C_{20}–C_{30}$ | candles, wax paper, etc. |
| | residue (asphaltic bitumen) | $C_{30}–C_{40}$ | asphalt, tar |

CHEMICAL PROPERTIES. Paraffins react with chlorine, under the influence of light, heat or catalysts, and to a lesser extent with bromine

$$C_nH_{2n+2} + Cl_2 \longrightarrow C_nH_{2n+1}Cl + HCl$$

In any paraffin the carbon atoms will have different numbers of hydrogen atoms attached. There are four types of carbon atom, with names corresponding to the number of bonds made *with other carbon atoms*:

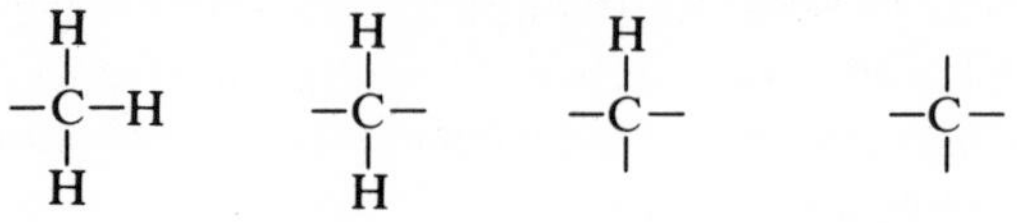

$$\overset{\displaystyle H}{\underset{\displaystyle H}{-\overset{|}{\underset{|}{C}}-H}} \qquad \overset{\displaystyle H}{\underset{\displaystyle H}{-\overset{|}{\underset{|}{C}}-}} \qquad \overset{\displaystyle H}{-\overset{|}{\underset{|}{C}}-} \qquad -\overset{|}{\underset{|}{C}}-$$

primary     secondary    tertiary    quaternary carbon

Chlorine removes hydrogen most easily from a tertiary carbon and least readily from a primary carbon. Where the first three types are present, a mixture will be produced:

$$3CH_3 \cdot CH_2 \cdot CH(CH_3)_2 + 3Cl_2 \longrightarrow \begin{array}{l} CH_3 \cdot CH_2 \cdot CCl(CH_3)_2 \\ CH_3 \cdot CHCl \cdot CH(CH_3)_2 + 3HCl \\ CH_2Cl \cdot CH_2CH(CH_3)_2 \end{array}$$

with further attack on other atoms possible. Well-known *chloro-paraffins* are:

|  | *Use* | *Boiling point* |
|---|---|---|
| $CH_2Cl_2$ methylene chloride | paint strippers | 40 °C |
| $CHCl_3$ chloroform | former anaesthetic; chemical intermediate | 61 °C |
| $CCl_4$ carbon tetrachloride | dry cleaner | 77 °C |

*Note:* the $CH_2<$ group is called a 'methylene' group.

Between 150 and 475 °C paraffins react as vapour with nitric acid e.g.

$$2CH_3 \cdot CH_2 \cdot CH_3 + 2HNO_3 \xrightarrow{\ 400\,°C\ }$$

$$CH_3 \cdot CH_2 \cdot CH_2 \cdot NO_2 + CH_3 \cdot CH(NO_2) \cdot CH_3 + 2H_2O$$

($+ CH_3 \cdot CH_2 \cdot NO_2 + CH_3 \cdot NO_2$ due to 'splitting' of paraffin). The products are called *nitroparaffins,* some of which are now finding use as paint solvents (see Table, 2, Chapter 9).

Paraffins burn, forming $CO_2$ and $H_2O$, hence their use as fuels.

## Olefins

In order to produce more petrol, the refiners pass some of the higher boiling fractions of petroleum through a heating process with either high pressure, or a catalyst. Large molecules are 'split' into smaller ones and the process is called *cracking*. As well as paraffins, compounds called *olefins* are

produced. These contain the group $\diagdown C=C\diagdown$ i.e. carbon atoms joined

by two bonds. The group is referred to as a 'double bond' or 'unsaturation'. A whole family of olefins exists, e.g.:

$CH_2{=}CH_2$    $CH{=}CH{-}CH_3$    $CH_2{=}CH{\cdot}CH_2{\cdot}CH_3$    $CH_3{\cdot}CH{=}CH{\cdot}CH_3$
ethylene     propylene          1-butylene           2-butylene

The prefix number indicates the *lower* numbered carbon atom involved in the double bond, numbering from one end of the molecule as with paraffins. Two or more double bonds in one molecule are possible, e.g.:

$$CH_2{=}CH{\cdot}CH{=}CH_2 \quad (1,3 \text{ butadiene})$$

Note the -ene ending, denoting an olefin, and the -di-, indicating the number of double bonds. Olefins with one double bond have the general formula, $C_nH_{2n}$.

Rotation is possible about a single bond, but not about a double bond. If the wooden balls and metal rods are used again to make the model, a double bond consists of two wooden balls joined by *two* metal rods. Even though the rods fit loosely in their sockets, it is not possible to twist the two balls in opposite directions if the rods are rigid. Thus

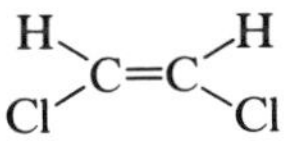

$$\underset{Cl}{\overset{H}{\diagdown}}C{=}C\underset{Cl}{\overset{H}{\diagup}} \quad \text{and} \quad \underset{Cl}{\overset{H}{\diagdown}}C{=}C\underset{H}{\overset{Cl}{\diagup}}$$

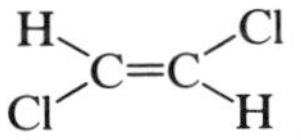

     *cis*-dichloroethylene     *trans*-dichloroethylene

are different compounds and are known as *geometrical isomers* of 1,2-di-chloroethylene.

The double bond is very reactive, the second bond breaking readily and the free valencies attaching themselves to other reactive substances, e.g.:

$$CH_2{=}CH_2 + \begin{array}{l} H_2 \\ Cl_2 \\ H_2O + [O] \\ \text{V. dilute } KMnO_4 \end{array} \begin{array}{l} \longrightarrow CH_3{\cdot}CH_3 \quad \text{ethane} \\ \longrightarrow CH_2Cl{\cdot}CH_2Cl \quad \text{ethylene dichloride} \\ \longrightarrow CH_2OH{\cdot}CH_2OH \quad \text{ethylene glycol} \end{array}$$

## Acetylenes

A third group of aliphatic hydrocarbons exists, called the *acetylenes*. Like acetylene itself ($HC{\equiv}CH$), all these compounds contain at least one *triple bond*. These are also unsaturated compounds and very reactive, but they are not found in the later chapters of this book.

## Terpenes

A special family of unsaturated compounds exists with the general formula $(C_5H_8)_n$, where $n = 2$ or more. These compounds are called *terpenes* and are composed of 2 or more molecules of *isoprene* bonded together

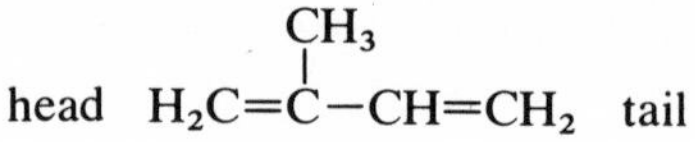

$$\text{head} \quad \underset{\text{isoprene}}{H_2C{=}\overset{\overset{\displaystyle CH_3}{|}}{C}{-}CH{=}CH_2} \quad \text{tail}$$

The isoprene molecules are usually (though not always) joined head-to-tail. The simplest terpenes (monoterpenes) are the chief constituents of the essential volatile oils obtained from the sap and tissues of certain plants and trees. These have been used in the manufacture of perfumes from earliest times. In particular, the monoterpenes extracted from pine trees, or obtained as by-products in paper pulp manufacture, have been used for many years as paint solvents (Chapter 9) and anti-oxidants (Chapter 10).

The solvents are mixed terpenes, but some of the main ingredients are:

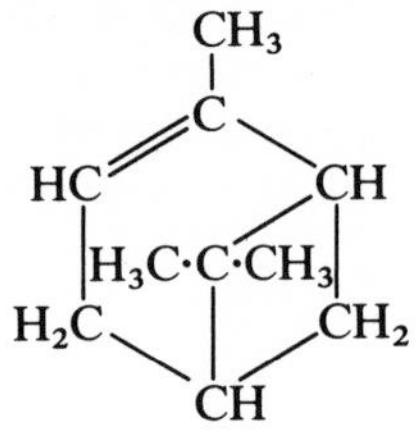

α-pinene
(turpentine)

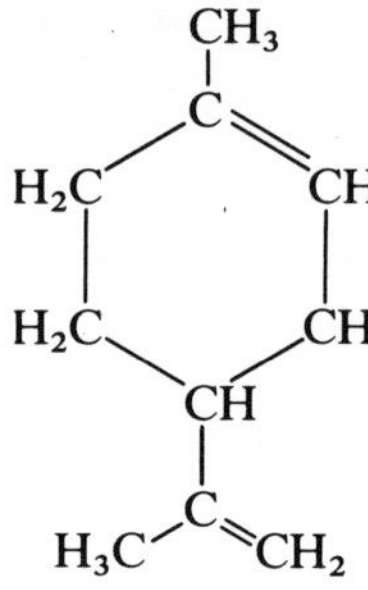

limonene
(dipentene)

Pine oil itself is a very complex mixture of alcohols (see below) derived from terpenes, e.g.

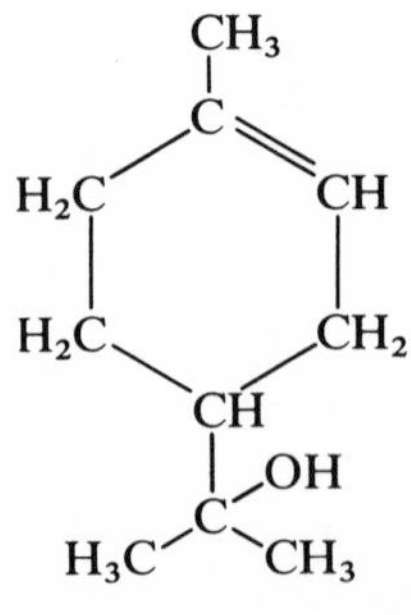

α-terpineol (note similarity to
limonene)

## Alcohols

If we imagine an aliphatic hydrocarbon, in which one hydrogen atom is removed and an —O—H group put in its place, we have an alcohol. The oxygen is bonded to carbon by a covalent bond and so, as distinct from a hydroxide ion, the group is called a *hydroxyl* group. A number of isomeric alcohols can be obtained, e.g. from the butanes:

| Paraffin isomer | Type of C atom chosen for substitution | Alcohol produced |
|---|---|---|
| (1) $CH_3 \cdot CH_2 \cdot CH_2 \cdot CH_3$ <br> *n*-butane | primary | $CH_3CH_2CH_2CH_2OH$ <br> *n*-butyl alcohol (butanol) |
| (2) *n*-butane | secondary | $CH_3 \cdot CH_2 \cdot CHOH \cdot CH_3$ <br> secondary (*sec*-) butyl alcohol |
| (3) $CH_3 \cdot CH(CH_3) \cdot CH_3$ <br> *iso*-butane | primary | $CH_3 \cdot CH(CH_3) \cdot CH_2OH$ <br> *iso*-butyl alcohol |
| (4) *iso*-butane | tertiary | $(CH_3)_3 \cdot C \cdot OH$ <br> tertiary (*ter*-) butyl alcohol |

There are thus three types of alcohol: primary ($—CH_2OH$), secondary ($>CHOH$) and tertiary ($\geqslant COH$), the name deriving from the type of carbon atom on which the theoretical substitution has been made.

The lower alcohols have fewer isomers:

| $CH_3OH$ | $CH_3 \cdot CH_2OH$ | $CH_3 \cdot CH_2 \cdot CH_2OH$ | $CH_3 \cdot CHOH \cdot CH_3$ |
|---|---|---|---|
| methyl alcohol | ethyl alcohol | *n*-propyl alcohol | sec-propyl alcohol |
| (methanol) | (ethanol) | (propanol) | (*iso*-propanol) |

There may, of course, be more than one hydroxyl group in the molecule:

TWO in a *glycol* (or dihydric alcohol), e.g., $\qquad CH_2OH \cdot CH_2OH$
ethylene glycol (anti-freeze)

THREE in a trihydric alcohol, e.g., $\qquad CH_2OH \cdot CHOH \cdot CH_2OH$
glycerine (glycerol)

FOUR in a tetrahydric alcohol, e.g.,
$$HOH_2C - \overset{\displaystyle CH_2OH}{\underset{\displaystyle CH_2OH}{C}} - CH_2OH$$

pentaerythritol

and so on. We will come across these polyhydric alcohols again in Part Two.

Methyl alcohol is prepared industrially by the reaction

$$CO + 2H_2 \xrightleftharpoons[\text{400 °C V. high pressure}]{\text{Zinc Chromite}} CH_3OH$$

Ethanol is what is usually meant by 'alcohol'. It is prepared by the fermentation of sugar using yeast. Alternatively, ethylene may be absorbed in concentrated sulphuric acid under pressure and at 75–80 °C:

$$CH_2{=}CH_2 + HO{\cdot}SO_2{\cdot}OH \longrightarrow CH_3{\cdot}CH_2{\cdot}O{\cdot}SO_2{\cdot}OH$$
ethyl sulphuric acid

Ethyl sulphuric acid is an ester (see below) of ethanol and the alcohol can be recovered when the ester is hydrolysed (see below) by dilution with water and the mixture is distilled.

## Carboxylic acids

Oxidation of primary alcohols produces a type of acid peculiar to organic chemistry, the carboxylic acid, e.g.

$$CH_3{\cdot}CH_2OH \xrightarrow{[O]} CH_3{\cdot}C{\nwarrow}^{H}_{O} \xrightarrow{[O]} CH_3{\cdot}C{\nwarrow}^{O-H}_{O} \quad (\text{or } CH_3{\cdot}COOH)$$

acetaldehyde        acetic acid
+ H$_2$O

An oxidizing agent, such as potassium permanganate, $KMnO_4$, is used rather than actual oxygen, hence the notation [O].

As with all types of organic compound, a large family of carboxylic acids, with different numbers and arrangements of carbon atoms, exists. All contain the carboxyl group —CO·OH.

They are acids in all senses of the word, e.g.

$$CH_3{\cdot}COOH \xleftarrow{\hspace{2cm}} CH_3{\cdot}COO^- + H^+$$

They are usually weak acids, though some, e.g. $CCl_3{\cdot}COOH$, are very strong.

Carboxylic acids may incorporate the features of organic compounds discussed previously in this chapter, e.g.

(i) An alkyl group, in $CH_3{\cdot}CH_2{\cdot}COOH$, propionic acid.
(ii) A double bond, in $CH_2{=}CH{\cdot}COOH$, acrylic acid.
(iii) A chlorine atom, in $CH_3{\cdot}CHCl{\cdot}COOH$, α-chloropropionic acid.

                 β       α

*Note:* The carbon atoms are labelled by Greek letters in alphabetical order, beginning next to the carboxyl group.

(iv) A hydroxyl group, in $CH_3{\cdot}CHOH{\cdot}COOH$, lactic acid (sour milk).

Just as alcohols can have more than one hydroxyl group, so acids can have more than one carboxyl group in the molecule, i.e. they may be *polybasic,* e.g. dibasic acids:

$$CH_2 \cdot COOH \qquad\qquad CH \cdot COOH$$
$$| \qquad\qquad\qquad || $$
$$CH_2 \cdot COOH \qquad\qquad CH \cdot COOH$$

succinic acid  maleic acid

The reader should not be put off because the organic molecules are becoming more complicated. On the whole, the reactions of a group (e.g. $>C{=}C<$, —OH, —COOH) are not affected by the complexity of the molecule. The main thing is to learn the important reactions of each group and apply them where possible.

The 'paraffinic' acids are known as the *fatty acids,* because the larger molecules of the series occur in oils and fats. The simpler members of the series are:

H·COOH  $CH_3 \cdot COOH$  $CH_3 \cdot CH_2COOH$  $CH_3 \cdot CH_2 \cdot CH_2COOH$
formic acid  acetic acid  propionic acid  butyric acid (smell of
(nettle stings)  (vinegar)   rancid butter)

Typical reactions of the carboxyl group are:

(1)  Salt formation with alkalis, e.g.

$$CH_3 \cdot COOH + NaOH \longrightarrow CH_3 \cdot COONa + H_2O$$
$$\text{sodium acetate}$$

If a higher fatty acid is used, a *soap* is formed, e.g. $C_{17} \cdot H_{35} \cdot COONa$, sodium stearate. Soaps are the oldest and simplest detergents and surfactants (see Chapter 10).

(2)

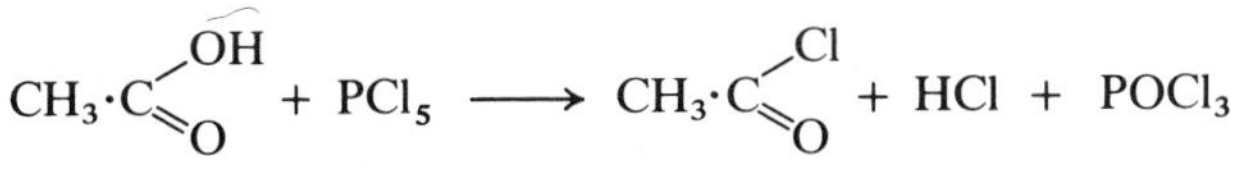

$$CH_3 \cdot C\!\!<^{OH}_{O} + PCl_5 \longrightarrow CH_3 \cdot C\!\!<^{Cl}_{O} + HCl + POCl_3$$

phosphorus  acetylchloride  phosphorus
pentachloride   oxychloride

This type of acid (or *acyl*) chloride is frequently used in reactions instead of the acid. It is more reactive than the acid and with water reforms the acid:

$$CH_3 \cdot COCl + H_2O \longrightarrow CH_3 \cdot COOH + HCl$$

(3) Salts and acid chlorides together give *acid anhydrides* (acids without water), e.g.

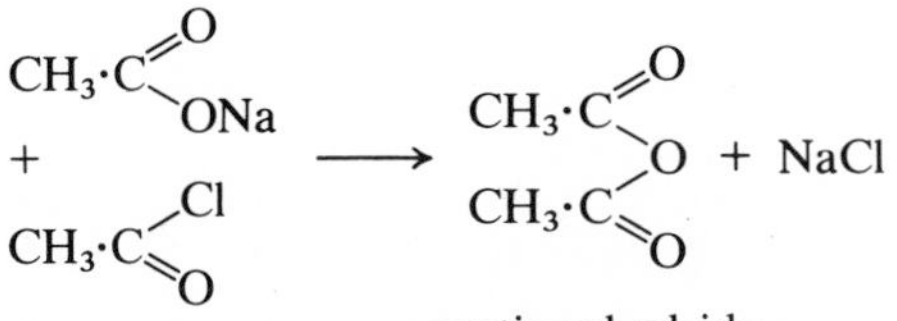

$$CH_3 \cdot C\!\!<^{O}_{ONa}$$
$$+ \qquad\qquad CH_3 \cdot C\!\!<^{O}_{O} + NaCl$$
$$CH_3 \cdot C\!\!<^{Cl}_{O} \qquad CH_3 \cdot C\!\!<^{O}$$

acetic anhydride

Anhydrides are also more reactive than acids and again are converted to acids by water:

$$(CH_3 \cdot CO)_2O + H_2O \underline{\qquad} 2CH_3 \cdot COOH$$

(4) Acids and alcohols react in a manner reminiscent of salt formation. The products are called *esters*, e.g.

$$CH_3 \cdot C{\overset{OH}{\underset{O}{\big\backslash}}} + C_2H_5 \cdot OH \xrightarrow[{[H_2SO_4]}]{heat} CH_3 \cdot C{\overset{O \cdot C_2H_5}{\underset{O}{\big\backslash}}} + H_2O$$

ethyl acetate

This can also be made by

$$CH_3 \cdot COCl + C_2H_5 \cdot OH \longrightarrow CH_3 \cdot CO \cdot OC_2H_5 + HCl$$

$$(CH_3 \cdot CO)_2O + 2C_2H_5OH \longrightarrow 2CH_3 \cdot CO \cdot OC_2H_5 + H_2O$$

## Esters

The whole range of acids and alcohols may be reacted to produce an enormous number of esters. They are found in a large number of natural and synthetic scents and perfumes. Even the simplest compounds, which are used as lacquer solvents (Chapters 9 and 11), have characteristic and generally pleasant smells. For example:

| | |
|---|---|
| ethyl acetate, $CH_3 \cdot CO \cdot C_2H_5$ | fruity |
| butyl acetate, $CH_3 \cdot CO \cdot O \cdot C_4H_9$ | pear drops |
| amyl acetate, $CH_3 \cdot CO \cdot O \cdot C_5H_{11}$ | pear drops |
| *iso*-amylacetate, $CH_3 \cdot CO \cdot O(CH_2)_2 \cdot CH(CH_3)_2$ | banana |
| *iso*-butyl propionate, $C_2H_5 \cdot CO \cdot O \cdot CH(CH_3) \cdot CH_2 \cdot CH_3$ | rum |
| butyl butyrate, $C_3H_7 \cdot CO \cdot O \cdot C_4H_9$ | pineapple |

Note that the ester names contain first the alcohol, then the acid portion

and end in -ate. All contain the ester linkage, $-\overset{\overset{\displaystyle O}{\|}}{C}-O-$

Esters can be converted back into the original acids and alcohols by *saponification* (soap formation) with alkali:

$$CH_3 \cdot CO \cdot O \cdot C_2H_5 + NaOH \longrightarrow CH_3 \cdot CO \cdot ONa + C_2H_5OH$$

$$\downarrow {\scriptstyle + HCl}$$

$$CH_3 \cdot CO \cdot OH + NaCl$$

or even by *hydrolysis* in boiling water

$$CH_3 \cdot COOC_2H_5 + H_2O \longrightarrow CH_3 \cdot COOH + C_2H_5OH$$

Because these reactions proceed so readily, the ester linkage between acids and alcohols is a relatively weak one and paint films containing ester linkages are attacked by bases generally (see Chapters 12 and 16).

Alcohols also form esters with inorganic acids, e.g.

$$C_2H_5OH + H_2SO_4 \longrightarrow C_2H_5O \cdot SO_2 \cdot OH + H_2O$$
ethyl sulphuric acid
(ethyl hydrogen sulphate)

and it is the reverse of this reaction (hydrolysis) that has already been quoted as a step in the manufacture of ethanol.

With hydrochloric acid

$$C_2H_5OH + HCl \longrightarrow C_2H_5 \cdot Cl + H_2O$$
chloroethane
(ethyl chloride)

and with nitric acid

$$C_2H_5OH + HNO_3 \longrightarrow C_2H_5 \cdot O \cdot NO_2 + H_2O$$
ethyl nitrate

Note the difference between this compound and the nitroparaffin, nitroethane:

$$CH_3 \cdot CH_2 \cdot N \overset{\displaystyle O}{\underset{\displaystyle O}{\diagup}} \qquad CH_3 \cdot CH_2 \cdot O \cdot N \overset{\displaystyle O}{\underset{\displaystyle O}{\diagup}}$$

  nitroethane                ethyl nitrate

Nitrate esters find wide use as explosives, e.g. nitroglycerin,

$$\begin{array}{l} CH_2 \cdot O \cdot NO_2 \\ | \\ CH \cdot O \cdot NO_2 \\ | \\ CH_2 \cdot O \cdot NO_2 \end{array}$$

Note that, although called *nitro*-glycerin, this compound is in fact glycerin *nitrate*. An equally misleading name is applied to another explosive that we will encounter in Chapter 11: nitrocellulose.

## Oils

The oils that are of particular interest to the paint chemist are the 'fatty' oils, which are largely vegetable oils pressed or extracted from the seeds or fruit of many types of vegetable matter. These are esters of glycerol, in which all three hydroxyl groups are reacted with fatty acids. The majority of these acids contain 18 carbon atoms. The triple esters are called *triglycerides*.

The usefulness of an oil is determined by the nature of the fatty acids present. Oils are classified according to their ability to dry to a solid film, when spread thinly and exposed to the air. *Drying oils* form a film at normal temperatures, *semi-drying oils* require heat, while non-drying oils will not form a film. More is said about the mechanism of the drying process in

| Oil | Fatty acids (%) | | | | | | |
|---|---|---|---|---|---|---|---|
| | Saturated acids | Oleic | Ricino-leic | Lino-leic | Lino-lenic | Eleo-stearic | Licanic |
| | No. of double bonds/acid molecule | | | | | | |
| | 0 | 1 | 1 | 2 | 3 | 3 | 3 |
| Drying | | | | | | | |
| Linseed | 10 | 22 | — | 17 | 51 | — | — |
| Perilla | 8 | 14 | — | 14 | 64 | — | — |
| Tung (Montana) | 5 | 9 | — | — | 15 | 71 | — |
| Oiticica | 12 | 7 | — | — | — | — | 81 |
| Semi-drying | | | | | | | |
| Soya bean | 13 | 28 | — | 54 | 5 | — | — |
| Tall | 7 | 45 | — | 48 | — | — | — |
| Safflower | 10 | 14 | — | 76 | — | — | — |
| Tobacco | 13 | 15 | — | 72 | — | — | — |
| Non-drying | | | | | | | |
| Castor | 10 | — | 87 | 3 | — | — | — |
| Coconut | 92 | 6 | — | 2 | — | — | — |

Chapter 12. Drying oils are found to contain a substantial proportion of fatty acids with three double bonds, semi-drying oils contain principally fatty acids with two double bonds, while non-drying oils contain only minor proportions of fatty acids with more than one double bond. The above table gives the fatty acid distributions of average samples of some common vegetable oils. Formulae of the unsaturated acids (all contain 18 carbon atoms):

Oleic $CH_3 \cdot (CH_2)_7 \cdot CH{=}CH \cdot (CH_2)_7 \cdot COOH$
Ricinoleic $CH_3 \cdot (CH_2)_4 \cdot CH_2 \cdot CHOH \cdot CH_2 \cdot CH{=}CH \cdot (CH_2)_7 \cdot COOH$
Linoleic $CH_3 \cdot (CH_2)_4 \cdot CH{=}CH \cdot CH_2 \cdot CH{=}CH \cdot (CH_2)_7 \cdot COOH$
Linolenic $CH_3 \cdot CH_2 \cdot CH{=}CH \cdot CH_2 \cdot CH{=}CH \cdot CH_2 \cdot CH{=}CH \cdot (CH_2)_7 \cdot COOH$
Eleostearic $CH_3 \cdot (CH_2)_3 \cdot CH{=}CH \cdot CH{=}CH \cdot CH{=}CH \cdot (CH_2)_7 \cdot COOH$
Licanic $CH_3 \cdot (CH_2)_3 \cdot CH{=}CH \cdot CH{=}CH \cdot CH{=}CH \cdot (CH_2)_4 \cdot CO \cdot (CH_2)_2 \cdot COOH$

Note that the double bonds may be *conjugated* (—CH=CH—CH=CH—), as in eleostearic and licanic acids, or *non-conjugated* (—CH=CH—CH₂— CH=CH—), as in linoleic and linolenic acids. Conjugated double bonds are directly joined by a single bond. Conjugated fatty acids give oils that dry faster and are less prone to yellowing (see p. 165). Fatty acids derived from oils (e.g. linseed oil fatty acids) can be obtained separately, if so desired. Catalytically treated versions of these fatty acids can also be obtained, in which the proportions of conjugated double bonds have been increased deliberately by an isomerization process.

Since each oil contains several acids in esterified form, one glycerol molecule is seldom esterified by three similar fatty acid molecules. It is not certain whether the distribution of fatty acids among the glycerol molecules is completely random, or influenced by some chemical factor to give a preponderance of particular arrangements. The exact distribution affects the drying considerably, since it affects the functionality of the molecules (see Chapters 5 and 12).

Castor oil is non-drying, but if water is removed from the ricinoleic acid by heating above 250 °C with sulphuric acid, a second double bond is introduced:

$$-CH_2-\underset{\underset{OH}{|}}{CH}-CH_2-CH=CH- \xrightarrow{-H_2O} \quad \begin{array}{l} -CH=CH-CH_2-CH=CH- \\ \text{non-conjugated (67–75\%)} \\[1em] -CH_2-CH=CH-CH=CH- \\ \text{conjugated (25–33\%)} \end{array}$$

The resulting *dehydrated castor oil* (*DCO*), although it contains no fatty acids with three double bonds, has such a high proportion of acids with two double bonds, including some conjugated unsaturation, that it is a drying oil. It does, however, dry to a film with a noticeable surface tack.

Castor oil can also be modified by reduction with hydrogen to produce *hydrogenated castor oil* (*HCO*)

$$-CH_2-\underset{\underset{OH}{|}}{CH}-CH_2-CH=CH- \xrightarrow{H_2} -CH_2-\underset{\underset{OH}{|}}{CH}-CH_2-CH_2-CH_2-$$

This has a special use as a paint additive (Chapter 10).

More will be said about oils and drying in Chapter 12.

# Four

# Organic chemistry: ethers to isocyanates

## Ethers

These compounds contain the characteristic ether linkage C—O—C, and can be thought of either as alcohols in which the hydrogen atom of the hydroxyl group is replaced by an alkyl group, or as 'alcohol anhydrides'. Diethyl ether and some other ethers are made by a dehydration reaction:

$$2C_2H_5OH \xrightarrow[140\,°C]{[H_2SO_4]} \underset{\text{diethyl ether}}{C_2H_5 \cdot O \cdot C_2H_5} + H_2O$$

This proceeds via the esterification of ethanol by sulphuric acid. Alternatively the dehydration can be carried out over aluminium oxide at 240–260 °C under high pressures.

In spite of this resemblance to anhydrides, ethers are very resistant to hydrolysis, and heating under pressure with dilute sulphuric acid is required to bring about

$$R—O—R + H_2O \longrightarrow 2ROH$$

The ether linkage is more readily broken by heating with halogen acids, particularly hydrogen iodide:

$$\underset{\text{methyl propyl ether}}{CH_3 \cdot O \cdot C_3H_7} + \underset{\substack{\text{hydrogen} \\ \text{iodine}}}{HI} \longrightarrow \underset{\substack{\text{methyl} \\ \text{iodide}}}{CH_3I} + \underset{\text{propanol}}{C_3H_7OH}$$

A variety of ethers with similar and dissimilar alkyl groups attached to the oxygen atom are possible and the ether linkage is found in many more complex compounds. Ethers are fairly unreactive compounds, though the hydrogen atoms of the methylene group next to the oxygen are more easily substituted by chlorine than the hydrogen atoms in a paraffin.

## Epoxides

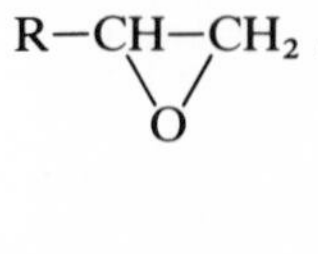

The oxide or epoxide ring, $R—\underset{\diagdown O \diagup}{CH}—CH_2$ , which, at first sight, appears to be a cyclic ether, is formed by direct reaction of ethylene and oxygen,

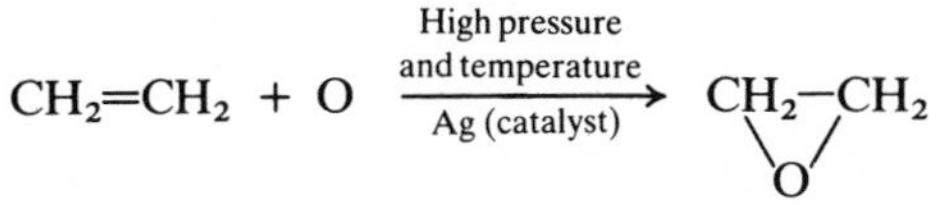

$$CH_2{=}CH_2 \ + \ O \ \xrightarrow[\text{Ag (catalyst)}]{\substack{\text{High pressure} \\ \text{and temperature}}} \ \underset{\underset{O}{\diagdown \diagup}}{CH_2{-}CH_2}$$

ethylene
oxide

or via a *chlorohydrin*, formed between an olefin and chlorine in alkaline solution

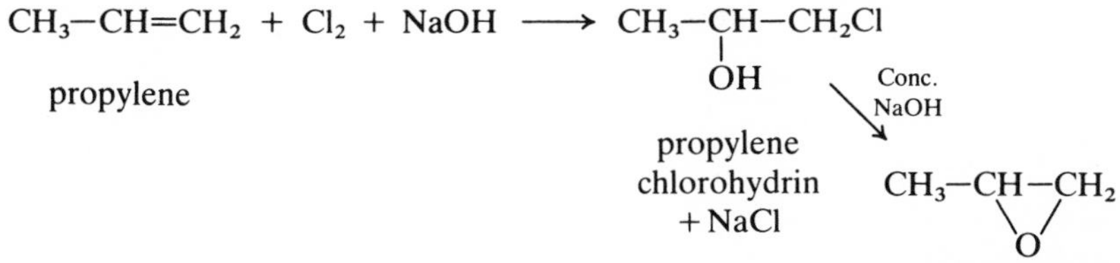

$$CH_3{-}CH{=}CH_2 \ + \ Cl_2 \ + \ NaOH \ \longrightarrow \ \underset{\underset{OH}{|}}{CH_3{-}CH{-}CH_2Cl}$$

propylene

propylene
chlorohydrin
+ NaCl

$$\xrightarrow[\text{NaOH}]{\text{Conc.}} \ \underset{\underset{O}{\diagdown \diagup}}{CH_3{-}CH{-}CH_2}$$

propylene oxide
+ NaCl + H$_2$O

The three membered ring is as reactive as a double bond and is readily opened by compounds containing an 'active' hydrogen atom (i.e. one attached to a strongly electronegative atom, such as N or O). In the ring-opening reaction, the hydrogen attaches itself to the ring's oxygen atom, e.g.

(1)

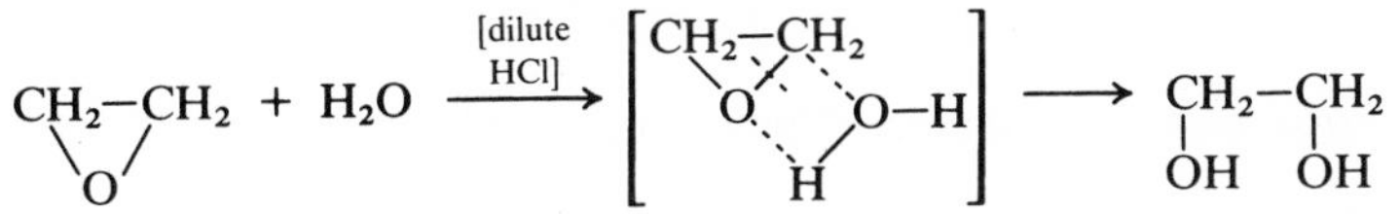

$$\underset{\underset{O}{\diagdown \diagup}}{CH_2{-}CH_2} \ + \ H_2O \ \xrightarrow[\text{HCl]}]{\text{[dilute}} \ \left[ \begin{array}{c} CH_2{-}CH_2 \\ \diagup \quad \diagdown \\ O \quad\ \ O{-}H \\ \diagdown\ \ \diagup \\ H \end{array} \right] \ \longrightarrow \ \underset{\underset{OH \ \ OH}{| \quad\ |}}{CH_2{-}CH_2}$$

ethylene oxide                                                  ethylene glycol

(2)

$$\underset{\underset{O}{\diagdown \diagup}}{CH_2{-}CH_2} \ + \ HOR \ \longrightarrow \ \underset{\underset{OH}{|}}{CH_2{\cdot}CH_2{\cdot}O{\cdot}R}$$

an ethylene glycol ether
(or ether-alchol, see Chapter 9)

where ROH is an alcohol

(3)

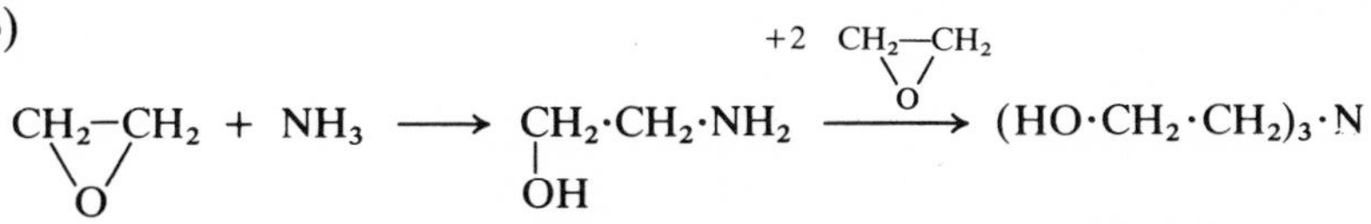

$$\underset{\underset{O}{\diagdown \diagup}}{CH_2{-}CH_2} \ + \ NH_3 \ \longrightarrow \ \underset{\underset{OH}{|}}{CH_2{\cdot}CH_2{\cdot}NH_2} \ \xrightarrow{+2 \ \underset{\underset{O}{\diagdown \diagup}}{CH_2{-}CH_2}} \ (HO{\cdot}CH_2{\cdot}CH_2)_3{\cdot}N$$

ethanolamine                                triethanolamine
(alcohol-amine)

(4)

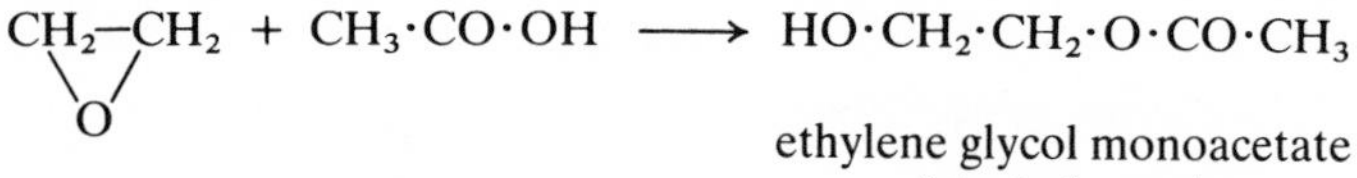

$$CH_2\!-\!CH_2 + CH_3\cdot CO\cdot OH \longrightarrow HO\cdot CH_2\cdot CH_2\cdot O\cdot CO\cdot CH_3$$
$$\underset{O}{\diagdown\diagup}$$

ethylene glycol monoacetate
(alcohol-ester)

This very reactive epoxide ring is the basis of the 'epoxy' paints.

### Aldehydes and ketones

We have already encountered the carbonyl group, $-\!\overset{\|}{\underset{O}{C}}-$ , as part of the

carboxyl group, $-C\overset{\diagup OH}{\diagdown_O}$ . It is also the basis of two other families of compounds, the aldehydes and ketones. These have the general formula, $R\!-\!\overset{\|}{\underset{O}{C}}\!-\!R'$ . R is an alkyl group. If R' is a hydrogen atom, the compound is an aldehyde, but if it is also an alkyl group, then the compound is a ketone. There is also the special case of formaldehyde where R and R' are both hydrogen atoms.

Thus the simplest aldehydes are

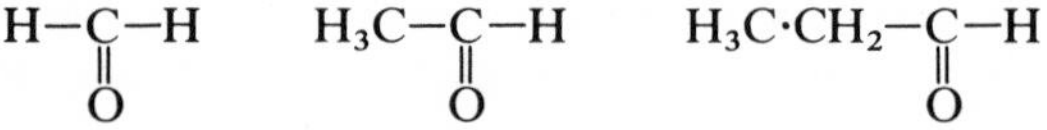

$$H\!-\!\overset{\|}{\underset{O}{C}}\!-\!H \qquad H_3C\!-\!\overset{\|}{\underset{O}{C}}\!-\!H \qquad H_3C\cdot CH_2\!-\!\overset{\|}{\underset{O}{C}}\!-\!H$$

     formaldehyde    acetaldehyde      propionaldehyde

The better known members of the ketone family are:

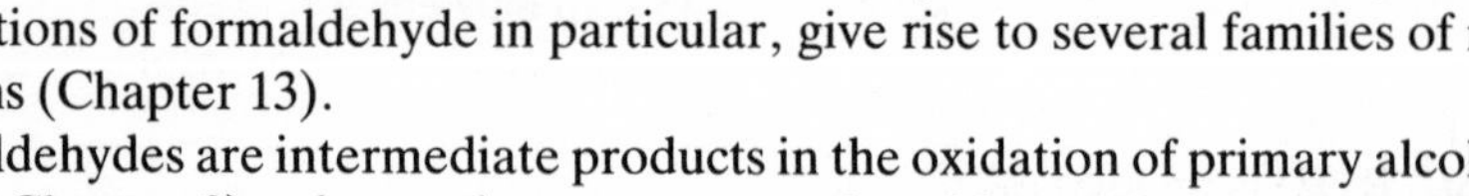

$$H_3C\!-\!\overset{\|}{\underset{O}{C}}\!-\!CH_3 \quad H_3C\!-\!\overset{\|}{\underset{O}{C}}\!-\!CH_2\cdot CH_3 \quad H_3C\!-\!\overset{\|}{\underset{O}{C}}\!-\!CH(CH_3)\cdot CH_2\cdot CH_3$$

    acetone      methyl ethyl ketone      methyl *iso*-butyl ketone

These ketones are all paint solvents and our interest in ketones is limited, in this book, to their performance as such. Aldehydes are not solvents, but the reactions of formaldehyde in particular, give rise to several families of new resins (Chapter 13).

Aldehydes are intermediate products in the oxidation of primary alcohols (see Chapter 3), whereas ketones are produced by oxidation of secondary alcohols

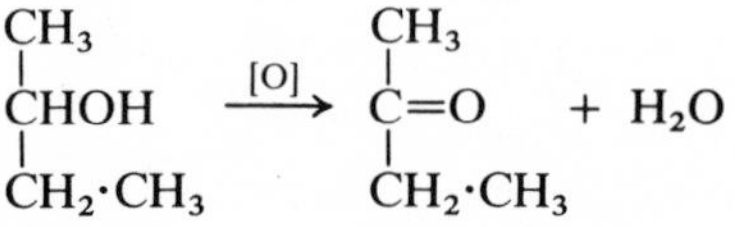

$$\begin{array}{ccc}
CH_3 & & CH_3 \\
| & \xrightarrow{[O]} & | \\
CHOH & & C\!=\!O \quad + H_2O \\
| & & | \\
CH_2\cdot CH_3 & & CH_2\cdot CH_3
\end{array}$$

    *iso*-butanol      methyl ethyl ketone

Although the reactions of the two families do differ at some points, they are largely the reactions of the carbonyl group, which readily accepts hydrogen from other compounds:

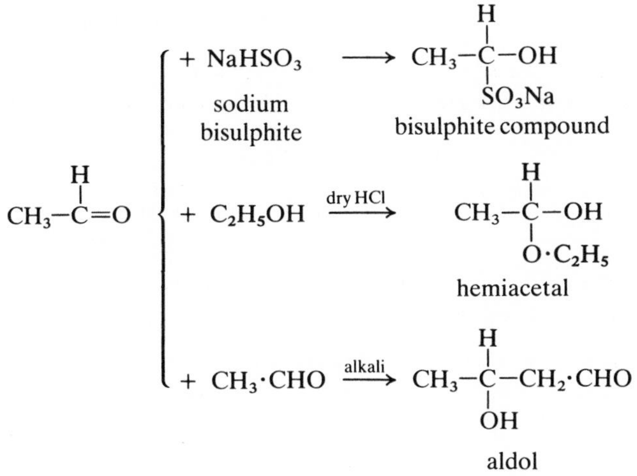

$$CH_3-\overset{\overset{\displaystyle H}{|}}{C}=O \left\{ \begin{array}{l} + NaHSO_3 \longrightarrow CH_3-\overset{\overset{\displaystyle H}{|}}{\underset{\underset{\displaystyle SO_3Na}{|}}{C}}-OH \\ \text{sodium} \quad\quad\quad \text{bisulphite compound} \\ \text{bisulphite} \\ \\ + C_2H_5OH \xrightarrow{\text{dry HCl}} CH_3-\overset{\overset{\displaystyle H}{|}}{\underset{\underset{\displaystyle O\cdot C_2H_5}{|}}{C}}-OH \\ \quad\quad\quad\quad\quad\quad\quad\quad \text{hemiacetal} \\ \\ + CH_3\cdot CHO \xrightarrow{\text{alkali}} CH_3-\overset{\overset{\displaystyle H}{|}}{\underset{\underset{\displaystyle OH}{|}}{C}}-CH_2\cdot CHO \\ \quad\quad\quad\quad\quad\quad\quad\quad\quad \text{aldol} \end{array} \right.$$

Sometimes, although the first step in the reaction may be addition of hydrogen to carbonyl, the overall reaction involves elimination of the carbonyl oxygen as water:

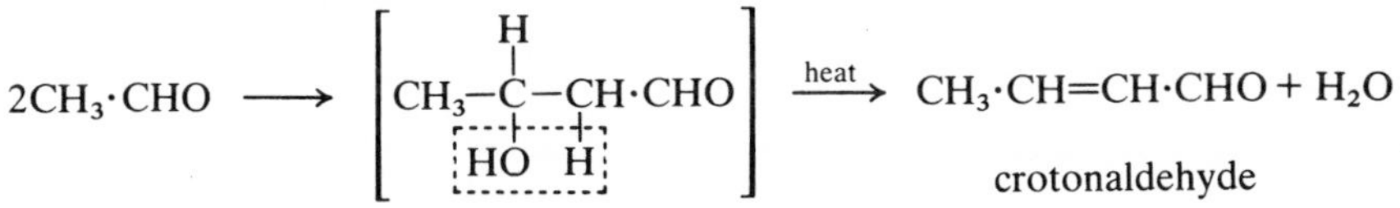

$$2CH_3\cdot CHO \longrightarrow \left[ CH_3-\overset{\overset{\displaystyle H}{|}}{\underset{\underset{\displaystyle \overline{HO \quad H}}{|}}{C}}-CH\cdot CHO \right] \xrightarrow{\text{heat}} CH_3\cdot CH=CH\cdot CHO + H_2O$$

crotonaldehyde

## Amines

Compounds containing carbon, hydrogen and nitrogen are introduced by a family of chemicals called amines. Amines are derived from ammonia and primary, secondary and tertiary forms are possible:

| $NH_3$ | $CH_3\cdot NH_2$ | $(CH_3)_2\cdot NH$ | $(CH_3)_3\cdot N$ |
|---|---|---|---|
| ammonia | methylamine | dimethylamine | trimethylamine |
| | (primary) | (secondary) | (tertiary) |

Like ammonia, these compounds are basic and therefore form salts with acids

$$NH_3 + HCl \longrightarrow NH_4Cl$$
$$CH_3\cdot NH_2 + HCl \longrightarrow CH_3\cdot NH_3Cl$$

$\left.\right\}$ nitrogen expanding its valency to 5

methylamine hydrochloride

The strength of the base increases with the number of alkyl groups attached to nitrogen. Amines can be prepared from ammonia and an alkyl halide:

$$C_2H_5Br + NH_3 \longrightarrow C_2H_5 \cdot NH_3Br \xrightarrow{+NaOH} C_2H_5NH_2 + NaBr$$
$$+ H_2O$$
ethylamine

But the $C_2H_5Br$ can react with the other hydrogen atoms in ethylamine, forming the secondary and the tertiary amine and finally a *quaternary ammonium salt*.

$$(C_2H_5)_3N + C_2H_5Br \longrightarrow (C_2H_5)_4 \overset{+}{N} \overset{-}{Br}$$
quaternary ethyl
ammonium bromide

Quaternary ammonium hydroxides are strong bases.

Amines react with epoxides in the manner of ammonia:

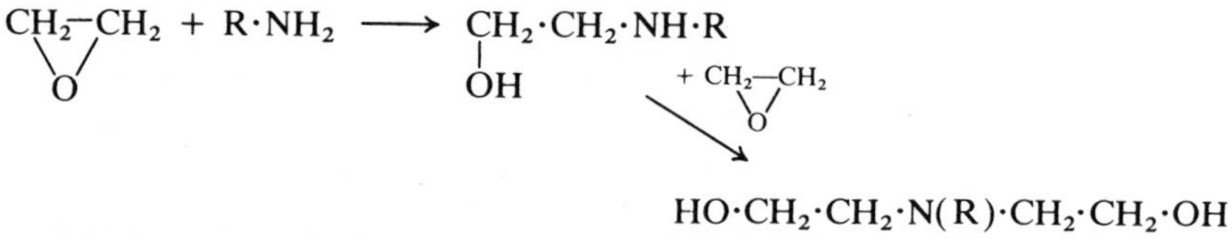

$$\underset{\underset{O}{\diagdown\diagup}}{CH_2-CH_2} + R \cdot NH_2 \longrightarrow \underset{\underset{OH}{|}}{CH_2 \cdot CH_2 \cdot NH \cdot R}$$

$+ CH_2-CH_2$ (epoxide)

$$HO \cdot CH_2 \cdot CH_2 \cdot N(R) \cdot CH_2 \cdot CH_2 \cdot OH$$

And with ketones, eliminating water

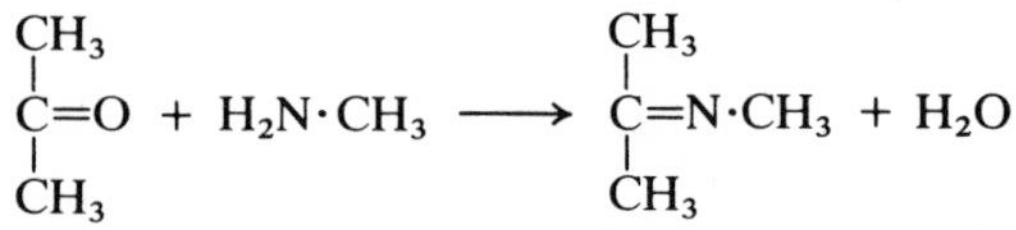

$$\underset{\underset{CH_3}{|}}{\overset{\overset{CH_3}{|}}{C}}=O + H_2N \cdot CH_3 \longrightarrow \underset{\underset{CH_3}{|}}{\overset{\overset{CH_3}{|}}{C}}=N \cdot CH_3 + H_2O$$

acetone                              a ketoxime

## Amides

Amides are neutral or mildly basic compounds containing the group, $-CO \cdot NH_2$. They are produced by reaction between acids, acid chlorides, acid anhydrides or esters and ammonia:

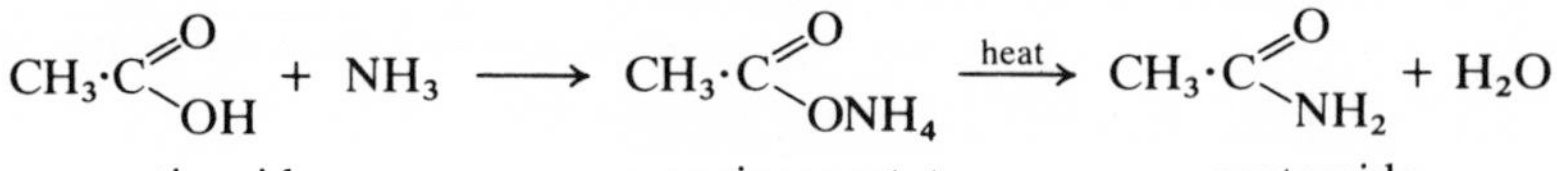

$$CH_3 \cdot C{\overset{O}{\diagdown OH}} + NH_3 \longrightarrow CH_3 \cdot C{\overset{O}{\diagdown ONH_4}} \xrightarrow{heat} CH_3 \cdot C{\overset{O}{\diagdown NH_2}} + H_2O$$
acetic acid                    ammonium acetate                    acetamide

$$CH_3 \cdot CO \cdot Cl + NH_3 \longrightarrow CH_3 \cdot CO \cdot NH_2 + HCl$$
acetyl
chloride

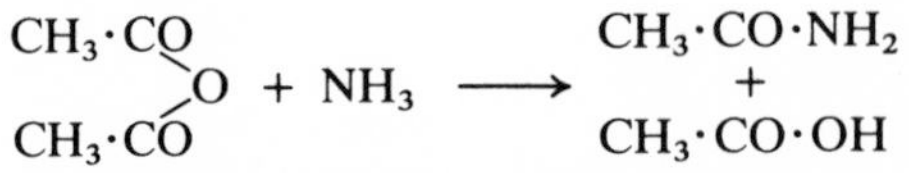

$$\underset{CH_3 \cdot CO}{\overset{CH_3 \cdot CO}{\diagdown\diagup}}O + NH_3 \longrightarrow \begin{matrix} CH_3 \cdot CO \cdot NH_2 \\ + \\ CH_3 \cdot CO \cdot OH \end{matrix}$$
acetic
anhydride

$$CH_3 \cdot CO \cdot O \cdot C_2H_5 + NH_3 \longrightarrow CH_3 \cdot CO \cdot NH_2 + C_2H_5OH$$

ethyl acetate

Substituted amides are formed by the reaction of amines with acid chlorides or anhydrides:

$$R \cdot NH_2 + (CH_3 \cdot CO)_2O \longrightarrow R \cdot NH \cdot CO \cdot CH_3 + CH_3 \cdot CO \cdot OH$$

N-substituted acetamide
R = alkyl

$$\begin{array}{c} R \\ \diagdown \\ R \diagup \end{array} NH + CH_3 \cdot CO \cdot Cl \longrightarrow \begin{array}{c} R \\ \diagdown \\ R \diagup \end{array} N \cdot CO \cdot CH_3 + HCl$$

di-N-substituted acetamide

The (—NH—CO—) amide linkage is found in polyamide resins.

Amides are resistant to hydrolysis, but this can be brought about by boiling with acetic and hydrochloric acids

$$R \cdot CO \cdot NH_2 + H_2O \longrightarrow R \cdot CO \cdot OH + NH_3$$

## Urea

Urea is the diamide of carbonic acid, $\begin{array}{c} HO \\ \diagdown \\ HO \diagup \end{array} C = O$ $(H_2CO_3)$ .

Phosgene may be considered to be the acid dichloride and it reacts with ammonia to give urea.

$$COCl_2 + 2NH_3 \longrightarrow H_2N \cdot CO \cdot NH_2 + 2HCl$$

phosgene                               urea

or alternatively it may be made directly from carbon dioxide

$$O=C=O + NH_3 \underset{\text{pressure}}{\overset{150\,°C}{\rightleftharpoons}} \left[ O=C \begin{array}{c} \diagup OH \\ \diagdown NH_2 \end{array} \right] \overset{NH_3}{\rightleftharpoons} O=C \begin{array}{c} \diagup ONH_4 \\ \diagdown NH_2 \end{array}$$

carbamic acid          ammonium
(unstable)             carbamate

$$O=C \begin{array}{c} \diagup NH_2 \\ \diagdown NH_2 \end{array} + H_2O$$

urea

Urea is the basic ingredient of urea formaldehyde resins.

## Aromatic compounds

These are an important series of compounds derived from benzene. The earliest known examples were marked by a spicy aroma, though the majority are not. Benzene is a liquid boiling at 80 °C. It is obtained by distillation of coal tar, one of the four products of the destructive distillation of coal.

The structure of benzene is the key to the distinctive properties of these compounds. The overall formula is $C_6H_6$ and the carbon atoms form a ring. However, the formula which correctly uses all the valencies, does not fit the reactions of the compound:

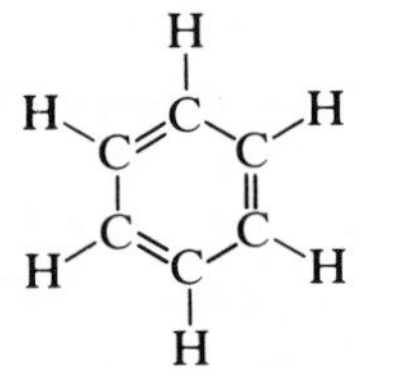

This formula suggests that benzene is an olefin and therefore various reagents should readily add themselves to the double bonds. In fact, only with difficulty can chlorine and hydrogen be made to react in this way to produce respectively $C_6H_6Cl_6$ (one form of which is the insecticide 'Gammexane') and $C_6H_{12}$ (*cyclo*-hexane), but it is relatively easy to *substitute* other atoms or groups for the hydrogen atoms. Compare the reactions of sulphuric acid with ethylene and benzene:

$$CH_2{=}CH_2 + H_2SO_4 \longrightarrow CH_3{\cdot}CH_2{\cdot}HSO_4$$
ethyl hydrogen sulphate

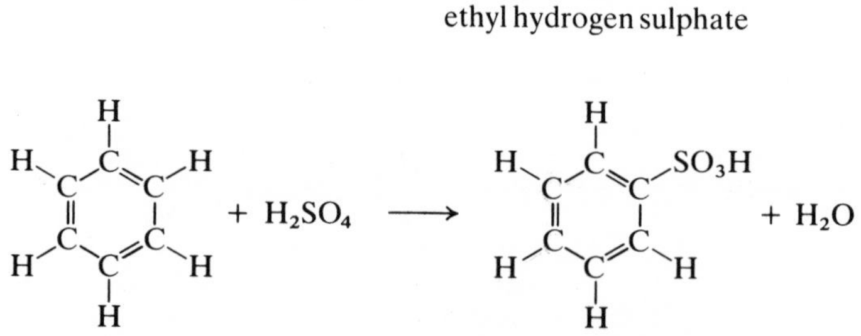

benzene sulphonic acid

Benzene, therefore, is not an ordinary olefin, yet it does have some olefin reactions. The above formula is not entirely satisfactory. Two more reasons can be given why this is so:

(1) If we substitute any two groups 'X' and 'Y' for the hydrogen atoms on

two adjacent carbon atoms, we should get two isomers:

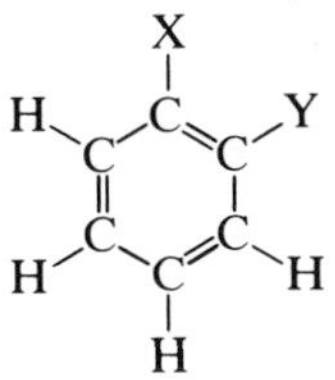

and

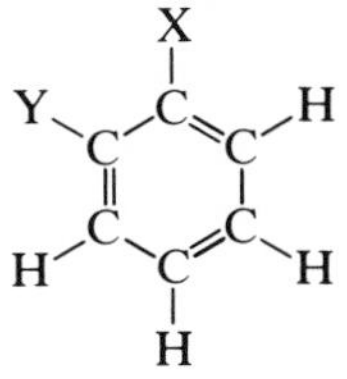

But these are never found.

(2) By the technique of X-ray crystallography, the distances between atoms can be measured. The distance between carbons (C—C) in ethane is 0·154 nm and in ethylene (C=C) it is 0·133 nm. Yet in benzene all adjacent carbon atoms are *equally* spaced 0·139 nm apart, a distance *between* the normal single and double bond lengths.

The full explanation of these facts is based on wave mechanics and the modern theories of electron distribution in chemical bonds. A simplified explanation is that the carbon atoms in benzene are held together by 1½ valencies (3 electrons). Two valency electrons are firmly located in a bond directly between each pair of carbon atoms, but the remaining electrons are 'pooled' between all six atoms and are free to move to particular atoms when the molecule is influenced by some outside reagent

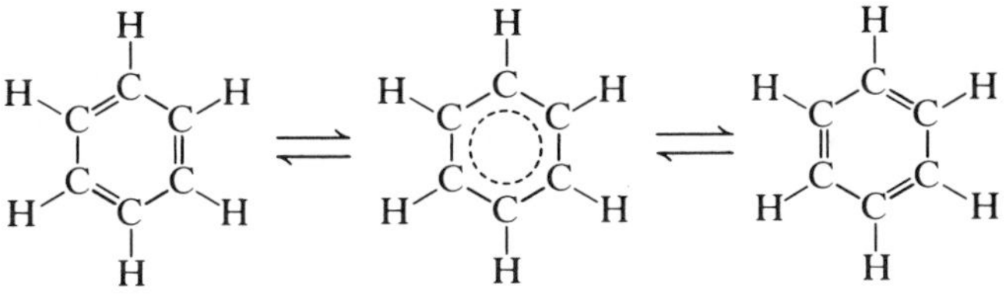

In practice, unless we wish to draw attention to a particular double bond or hydrogen atom, we depict the formula as a hexagon,  to which substituent groups may be added, e.g. benzene sulphonic acid,

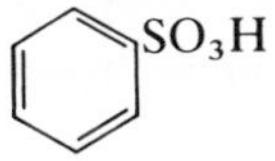

Benzene is a toxic chemical, causing anaemia and cancer if inhaled in prolonged doses. For this reason, it is not used as a paint solvent.

**Benzene substitution products**

Since some or all of the six hydrogen atoms of benzene may be replaced by other atoms or groups, a large variety of aromatic compounds is possible and many of these will be isomers. Isomers are distinguished from one another by numbering the carbon atoms from 1 to 6, so that the atoms on which substitution has occurred can be clearly defined.

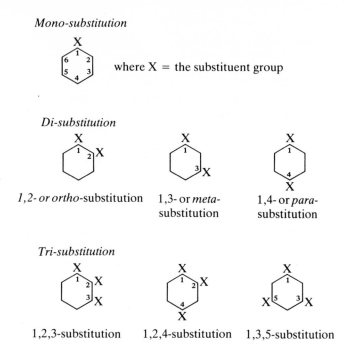

*Mono-substitution*

where X = the substituent group

*Di-substitution*

*1,2- or ortho*-substitution     1,3- or *meta*-substitution     1,4- or *para*-substitution

*Tri-substitution*

1,2,3-substitution     1,2,4-substitution     1,3,5-substitution

and so on.

Note that the first group is always given position 1 and the others are given the lowest numbers possible.

Every type of organic compound covered in the aliphatic series is also possible in the aromatic series: hydrocarbons, alcohols, acids, esters, ethers, aldehydes, ketones, amines, amides and others. On the whole, groups behave exactly as they do in aliphatic compounds, but sometimes they have their behaviour slightly modified by the direct attachment to the benzene ring. Aromatic alcohols, for example, are weakly acid in character

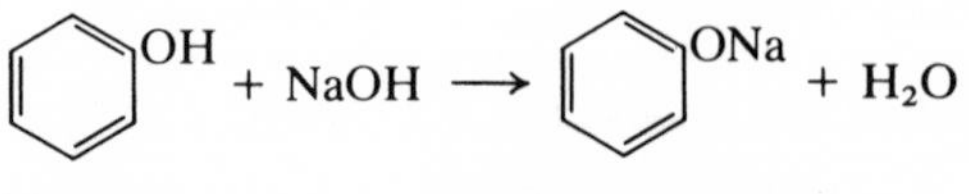

$$\text{phenol}\ \ \text{OH} + NaOH \longrightarrow \text{sodium phenoxide}\ \ \text{ONa} + H_2O$$

They do not give esters with carboxylic acids, but will do so with acid chlorides. They differ markedly from alcohols in the ease with which substitution takes place (in the 2 and 4 positions), e.g.

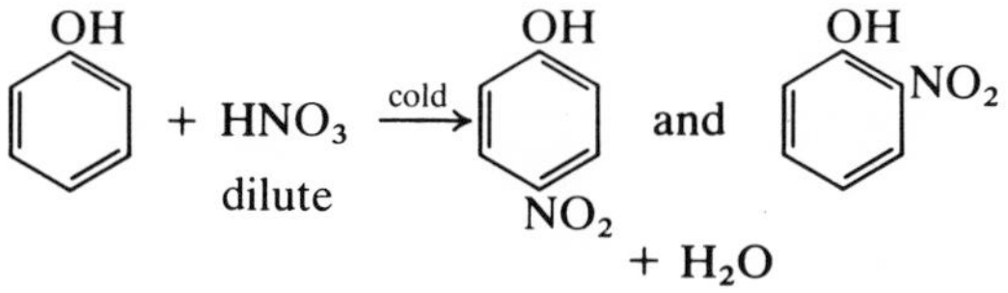

$$+ H_2O$$

*p* (para)    *o*-(ortho)
nitrophenol   nitrophenol

Nitrates are not formed. Because of these differences, the family of compounds is called *phenols,* not alcohols.

The following are aromatic compounds that will be encountered later in the book.

*HYDROCARBONS*

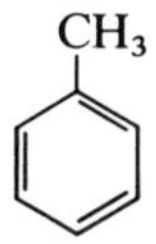

toluene

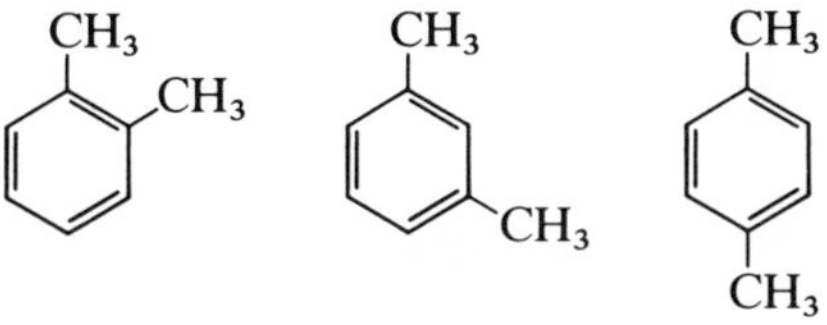

*o*-_____*m*-_____*p*-

xylene

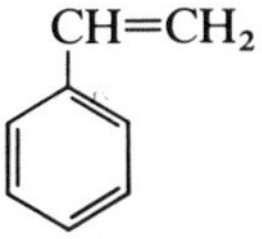

Styrene
(vinyl benzene)

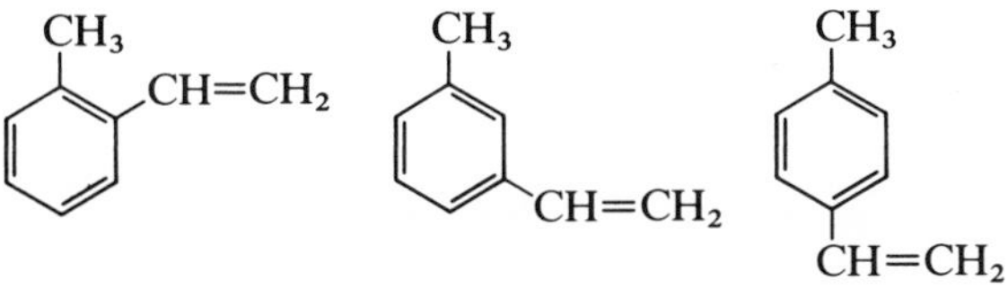

*o*-_____*m*-_____*p*-

vinyl toluene

The aliphatic hydrocarbon groups attached behave in a typical manner.

*ACIDS*

benzoic acid

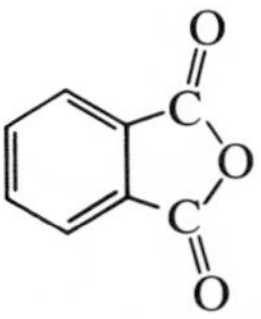

*o*-phthalic anhydride

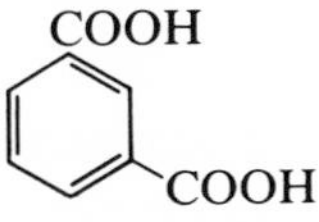

*m*- or *iso*-phthalic acid

*ALCOHOL*

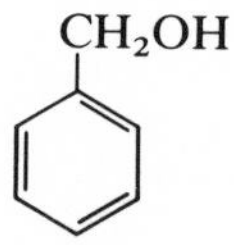

$$CH_2OH$$

benzyl alcohol

This is an aromatic-aliphatic compound. As the hydroxyl group is attached to the aliphatic portion, this is a true alcohol, not a phenol.

*ESTER*

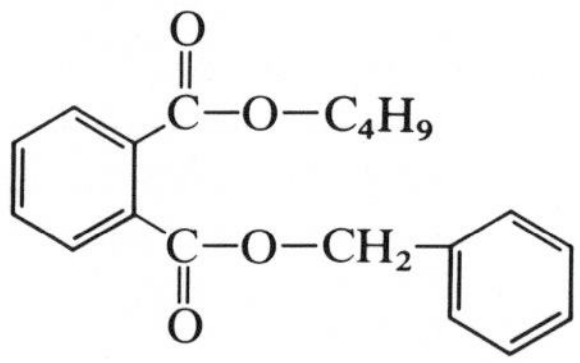

butyl benzyl phthalate

*AMINE*

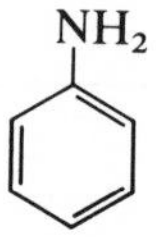

$$NH_2$$

aniline

## Isocyanates

The isocyanates are a family of compounds existing in aliphatic and aromatic forms, but used chiefly in the latter. All isocyanates contain the isocyanate group: —N=C=O, e.g.

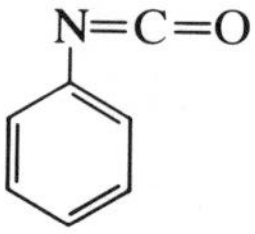

$$N=C=O$$

phenyl isocyanate

Note that the phenyl group is — or $C_6H_5$—.

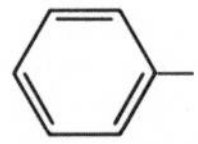

Isocyanates are prepared by the following route

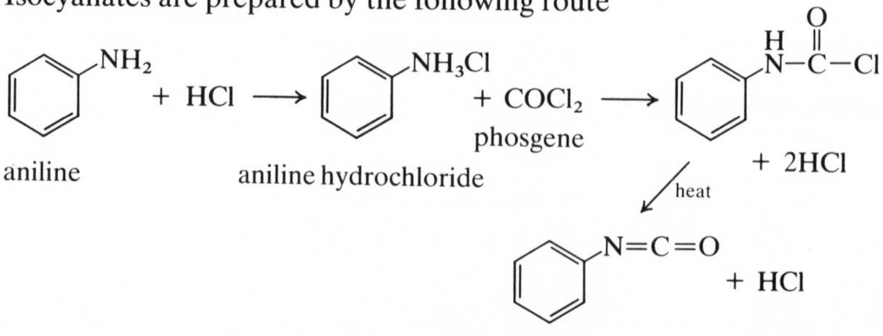

aniline                    aniline hydrochloride                    + 2HCl

phosgene

+ HCl

phenylisocyanate

Two or more isocyanate groups may be included in a molecule. Some well-known compounds are:

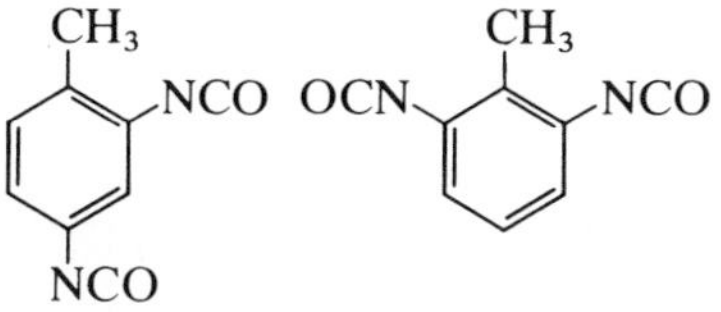
CH3
NCO
OCN
CH3
NCO
NCO

2,4- and 2,6-
toluene diisocyanates

$OCN{\cdot}(CH_2)_6{\cdot}NCO$

hexamethylene diisocyanate

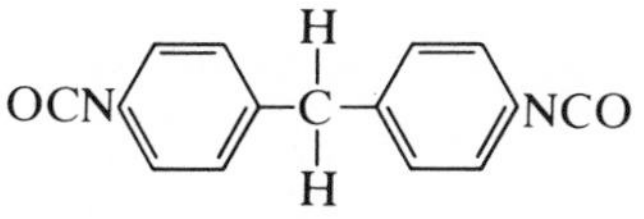
H
OCN
C
NCO
H

diphenyl methane-4,4′-diisocyanate
(4′ implies substitution in the second ring)

The reactions of the isocyanate group occur with compounds containing an 'active' hydrogen atom. In each example, the hydrogen atom attaches itself to nitrogen in the —NCO group, opening the N=C double bond. The remainder of the reacting molecule then attaches itself to the spare valency which has been made available on the —NCO carbon. Thus

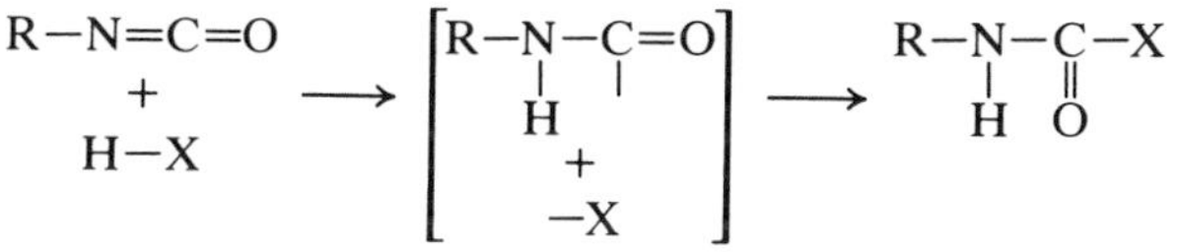
R–N=C=O
+
H–X
R–N–C=O
H
+
–X
R–N–C–X
H O

In the following examples R is an alkyl or aromatic (aryl) group and R′ is some other alkyl or aryl group.

(1) 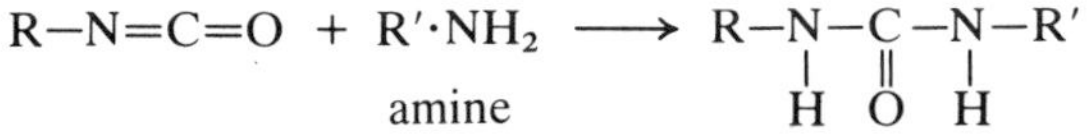
R–N=C=O + R′·NH2
amine
R–N–C–N–R′
H O H

a substituted urea

(2) 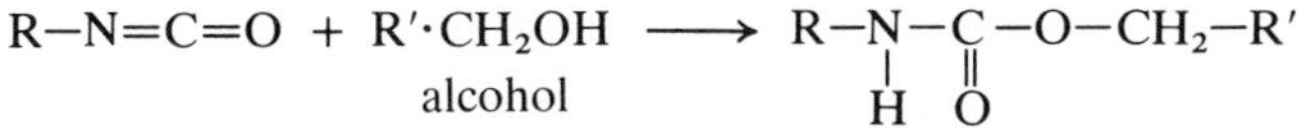
R–N=C=O + R′·CH2OH
alcohol
R–N–C–O–CH2–R′
H O

secondary and tertiary alcohols are less reactive.

a urethane

(3) 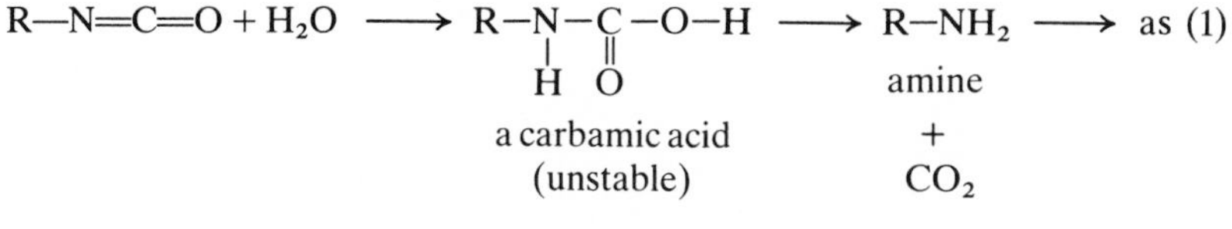
R–N=C=O + H2O
R–N–C–O–H
H O
a carbamic acid
(unstable)
R–NH2
amine
+
CO2
as (1)

(4) $R\text{—}N\text{=}C\text{=}O + R\cdot COOH \longrightarrow$
$$R\text{—}\underset{H}{\overset{|}{N}}\text{—}\underset{\overset{\|}{O}}{C}\text{—}O\text{—}\underset{\overset{\|}{O}}{C}\text{—}R' \longrightarrow R\text{—}\underset{H}{\overset{|}{N}}\text{—}\underset{\overset{\|}{O}}{C}\text{—}R'$$

some acids          an anhydride of      an amide
       a carbamic acid        $+ CO_2$
       (unstable)

Reaction (1) proceeds the most readily and at the lowest temperature. Each subsequent reaction occurs a little less readily and at a slightly higher temperature going from (1) to (4), but all proceed at normal atmospheric temperatures. Further reactions can proceed with the 'active' hydrogens in the urea, urethane and amide products, but these are less important, except at paint stoving temperatures (100 °C and above).

# Five

# Solid forms

This chapter is concerned with the *arrangement* of molecules in solids.

### Crystalline solids

Pure, simple substances separate from solution on cooling as *crystals*. These crystals have a sharp melting point, which marks precisely the boundary between the liquid and solid states. They also have a definite and consistent shape. This shape is due to the pattern in which the atoms, ions or molecules are arranged. Thus crystals of common salt are always cubic. The arrangement of the $Na^+$ and $Cl^-$ ions is in a cubic pattern (Fig. 8), which extends in all directions to fill and form the shape of a single crystal.

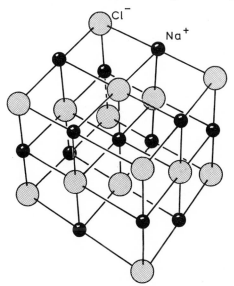

**Fig. 8** Arrangement of ions in a crystal of common salt.

Other inorganic compounds have a variety of crystal shapes, in all of which the ions, atoms or molecules are arranged in a definite pattern, held together by ionic or covalent bonds, or by intermolecular attractions, which arise from the polarity (or electrical asymmetry) of the molecules. In diamond, for example, each carbon atom is covalently bonded to four neighbours (Fig. 9) so that the crystal is one 'giant molecule'. This in part accounts for its great strength and hardness.

Organic crystals contain more complex molecules arranged in patterns decided by the molecular shape and the intermolecular attractions. Thus urea has the tetragonal crystal shape (Fig. 10), with molecules arranged as shown in Fig. 11, each amino group being directed towards a carbonyl oxygen atom, so that N—H··O hydrogen bonding takes place.

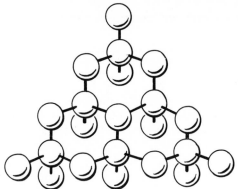

**Fig. 9** The structure of a diamond crystal.

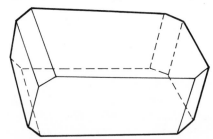

**Fig. 10** Tetragonal shape of a urea crystal.

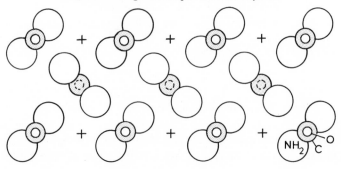

**Fig. 11** Arrangement of molecules in a urea crystal.

## Amorphous solids

In many solids, the molecules are not arranged in any pattern at all. Instead they contain the random arrangement of molecules typical of a liquid, which appears to have become 'fixed' as a solid by cooling. Since there is no pattern of molecules, there is no typical shape to the solid, which is known as an *amorphous* (shapeless) *solid.*

The act of crystallization is a sudden process, which occurs at the precise temperature at which the energy of movement yields to the forces of attraction. In an amorphous solid, the molecules gradually become slower and more closely attracted to one another, until they are so closely packed that they completely impede one another and the substance will not flow. It is impossible to say *exactly* when this occurs, or even whether the substance has changed from flowing very slowly indeed, to not flowing at all. There is no melting point and the boundary between the solid and the liquid state is blurred.

As the amorphous solid is warmed, the energy it receives is, at first, only sufficient to allow some movement of the less impeded portions of the large molecules that usually form these solids. Freedom of movement spreads to other parts of the molecules as heating continues until whole molecules can move. At this stage, the substance is obviously in the liquid state.

The temperature at which the molecules begin to obtain partial freedom of movement, is known as the *glass* (or second order transition) *temperature*. This can be determined by plotting a graph of, for example, the density of the substance against temperature. There is a sharp change in the slope of the plot at the glass temperature.

Various *softening temperatures* are often quoted for these substances. These correspond approximately to the state of obvious fluidity, but vary appreciably with the apparatus and method used for measurement.

Examples of amorphous solids are glass, tar and the naturally occurring gums and resins. Such resins formed the basis of many early paints, because the distinctive property of an amorphous solid is that, when a melt is poured into a tray, or when a solution of the solid is poured upon a surface and the solvent is allowed to evaporate, the resultant solid forms a *continuous* film. Because it has no natural shape of its own, it takes the shape into which it has been cast. A crystalline material would give a discontinuous film, consisting of many hundreds of tiny individual crystals. It is the amorphous continuity that makes the resin film extremely suitable for protecting surfaces.

## Polymers

Natural resins are complex mixtures of different substances with fairly large molecules. Nowadays many synthetic resins are used in paints. These resins have many properties that their natural predecessors had not, but they also consist of mixtures of different molecules, all of which are large.

Synthetic resins are *polymers*. 'Poly-mer' means 'many parts'. A polymer molecule is composed of many smaller parts, contributed by similar or dissimilar simple molecules, which are joined together until there are hundreds or thousands of atoms in the polymer molecule.

### Addition polymers

For example, ethylene, $CH_2{=}CH_2$, is a simple molecule. If ethylene gas is highly compressed (20 000 p.s.i.), heated to 200 °C and allowed to come into contact with a carefully controlled trace of oxygen, the following occurs:

(1) Oxygen attacks a double bond, opening it. Temporarily a free radical has been created

$$O_2 + CH_2{=}CH_2 \longrightarrow \underset{\underset{\overset{|}{O.}}{\overset{|}{O}}}{CH_2}{-}\overset{\bullet}{C}H_2 \quad (\text{or } R{-}CH_2\cdot)$$

(2) The highly reactive free radical attacks another double bond

$$\underset{\underset{R}{\overset{|}{C}H_2}}{\overset{\bullet}{C}H_2{=}CH_2} \; + \; \longrightarrow \; \underset{\underset{R}{\overset{|}{C}H_2}}{\overset{\overset{|}{C}H_2{-}CH_2}{}} \; \xrightarrow{+\,CH_2{=}CH_2} \; \underset{\underset{R}{\overset{|}{C}H_2}}{\overset{\overset{\overset{|}{C}H_2{-}\overset{|}{C}H_2}{CH_2{-}CH_2}}{}} \quad \text{and so on.}$$

A *chain reaction* has been set in motion and the $-CH_2-CH_2-$ unit will repeat itself a thousand times or more in forming a huge molecule of the polymer, polyethylene ('Polythene').

Polyethylene is not used in paints, because of its marked crystallinity (see below). However, many compounds containing the vinyl ($CH_2{=}CH-$) or vinylidene ($CH_2{=}C{<}$) group, can be polymerized by a free radical route and many of the products are useful in paints, e.g.

$CH_2{=}CHCl \longrightarrow$ PVC (polyvinyl chloride)
vinyl chloride

$CH_2{=}CH{\cdot}O{\cdot}CO{\cdot}CH_3 \longrightarrow$ PVAc (polyvinyl acetate)
vinyl acetate

$CH_2{=}C(CH_3){\cdot}CO{\cdot}O{\cdot}CH_3 \longrightarrow$ Polymethyl methacrylate ('Perspex')
methyl methacrylate

Usually the *monomer* (as the starting compound is called) is liquid and can be polymerized in bulk or in solution. The *initiator* of polymerization is not oxygen, but some compound which, on heating, decomposes into frree radicals (e.g. organic peroxides, Chapter 16). The growth of the polymer chain is terminated by one of the following occurrences:

(1) *Combination*. An encounter between two polymer free radicals which satisfy one another's free valencies:

$$R\cdot + \cdot R' \longrightarrow R—R'$$

(2) *Disproportionation*. A similar encounter, in which one free radical removes a hydrogen atom from the other to become a saturated compound. The double radical rearranges to an unsaturated compound; neither free radical survives:

$$\begin{matrix} R'—CH—CH_2\cdot & & R'—CH—CH_2\cdot & & R'—CH=CH_2 \\ | & & & & \\ H & \longrightarrow & + & \longrightarrow & + \\ & & RH & & RH \\ +\, \dot{R} & & & & \end{matrix}$$

(3) *Chain transfer*. The polymer free radical is satisfied by the removal of a monovalent atom (usually hydrogen) from another molecule:

$$R\cdot + R'H \longrightarrow RH + R'\cdot$$

One polymer chain is ended, but another begins, as a result of the creation of a new free radical. Hence the chain reaction is 'transferred' to another molecule. The hydrogen atom may be removed from

(a) another monomer molecule
(b) a solvent molecule
(c) a compound called a *chain transfer agent*, specially included in the preparation for this purpose.

It is already obvious that the molecules produced are large. They will also be different, even if only one monomer is used, because the chemist cannot control the size of each individual molecule. He must leave this to chance encounters between molecules in the reaction mixture. Any sample of polymer will contain a range of molecular sizes, some chains being terminated early and others later in their growth. The *average* size can be controlled by the amounts of initiator and chain transfer agent used (larger amounts of both giving smaller molecules) and by variations in the practical conditions under which the polymerization is performed.

The polymer molecular size is described by its *molecular weight*. If a hydrogen atom is given a weight of one, carbon, which is approximately twelve times heavier, is 12, nitrogen, 14, oxygen, 16, and so on. The molecular weight is found by adding together the relative weights (called *atomic weights*) of all the atoms in the molecule. A polymer molecular weight might vary from about 1000 to over a million. By contrast, ethylene has a molecular weight of only 28.

Each of the polymers listed above is made from one monomer only. These polymers are called *homopolymers*. Any number of coreactive vinyl or vinylidene monomers can be polymerized together and the product is called a *random copolymer*. In a copolymer molecule, the different monomer units are combined together in a random way. As the polymer free radical grows, the next unit to be added is decided by chance: it depends upon which monomer molecule encounters the right part of the chain first

$$A + B \rightarrow A.A.A.B.B.A.B.B.A.A.B.A.B.B.B.B. \text{ etc.}$$

Individual molecules will differ in the proportions of the comonomer units, but the overall proportions will depend on the ratio of the monomers charged into the reaction vessel. Copolymers have properties in between those of the homopolymers which could be made from the separate components. Thus polymethyl methacrylate is hard, relatively brittle and insoluble in petrol. Polybutyl methacrylate is softer, more flexible and soluble in petrol. A copolymer, predominantly methyl methacrylate, could remain insoluble in petrol (though softened by it) and yet be more flexible than polymethyl methacrylate.

Sometimes it is possible, by carrying out the polymerisation in stages, to arrange the monomer units in runs or blocks, e.g.

$$...A.A.A.A.A.A.B.B.B.B.B.B... \text{ etc.}$$

These more orderly copolymers are known as *block copolymers*. An A–B block has molecules with a chain of type A monomer units joined at one end to a chain of type B units. A–B–A blocks have only type A chains at the ends of molecules. A–B–A block copolymers can be used as associative thickeners (see Chapter 10).

## Condensation polymers

The polymerizations described so far have all been of one type: *addition polymerization*. The polymer molecule grows by the addition of another monomer unit *without the loss of any atoms* in the process. The alternative type of polymerization is known as *condensation polymerization*. Here each growth reaction is accompanied by the production of a small molecule (usually water) which is excluded from the polymer chain, e.g.

$$HO \cdot OC \cdot R \cdot CO \cdot OH + HO \cdot R' \cdot OH \longrightarrow HO \cdot OC \cdot R \cdot CO \cdot O \cdot R' \cdot OH + H_2O$$
dibasic acid       dihydric alcohol

The esterification of one group from each molecule produces water and the product on the right, containing a carboxyl group at one end and a hydroxyl group at the other, is the repeating unit (or monomer) for the polymerization. This proceeds as follows:

HO·R'·OH + HO·OC·R·CO·O·R'·OH + HO·OC·R·CO·OH ⟶

HO·R'·O·OC·R·CO·O·R'·O·OC·R·CO·OH + 2H₂O, and so on

Each reaction produces another ester group and the resultant polymer is a polyester resin.

Esterification need not be the reaction concerned; polyamides, epoxy resins, amino resins and others are produced by condensation polymerization. It is worth mentioning that, in condensation polymerizations, the by-product is often an embarrassment and must be removed before the resin can be used. Water, for example, is immiscible with polyester resins and can be removed by distillation during the polymerization. Removal of the by-product can accelerate the polymerization.

Addition and condensation polymerizations take very different courses. In addition polymerizations, the highly reactive nature of a free radical means that, once one is formed, it will grow by chain reaction to termination in a matter of seconds. The polymerization proceeds by the addition of single small monomer molecules to a large growing chain. Completed polymer chains and unreacted monomer exist together until polymerization is 100 per cent. The total time for an addition polymerization, which may be a matter of hours, depends upon the rate of formation of new free radicals and the yield of polymer required. Under the right conditions, very high molecular weights (greater than 100 000) are easily obtained.

Condensation polymerizations, on the other hand, depend on ordinary reactions between reactive organic groups. There is no chain reaction. All the original monomer molecules soon double their size and further growth depends upon the joining together of larger and still larger units. The concentration of reactive groups falls with each reaction and the viscosity rises so that, towards the end of the polymerization, the rate of reaction is very slow indeed, since the right groups have difficulty in finding one another. Consequently, although the yield of low molecular weight polymer is 100 per cent after an hour or two, the attainment of molecular weights as high as 20 000 is very difficult and, even if it can be done, it involves very long reaction times. The total time for a condensation polymerization therefore depends upon the molecular weight required and not upon the yield.

## Functionality

It should be noted that a molecule with the ability to react with only one other molecule, cannot produce a polymer. Let us assume that X and Y are reactive groups and that X will react with Y, but two Xs and two Ys will not combine. Then

R—X + Y—R' ⟶ R—XY—R'

and no further reaction is possible since X and Y are no longer reactive. Also

$$R—X + Y—R'—Y + X—R \longrightarrow R—XY—R'—YX—R$$

Unless Y can react with Y, no polymer can be obtained.

The useful monomer must have the ability to react with a minimum of two other molecules. In condensation polymerization, this means two reactive groups per molecule; in addition polymerization, a double bond has the ability to produce two free valencies:

$$CH_2{=}CH_2 \longrightarrow —CH_2—CH_2—$$

The number of molecules with which a monomer molecule may combine in the polymerization reaction concerned, is known as its *functionality*. If the monomers present have a functionality of *two*, a linear polymer is produced:

—A—A—A—A—A..., etc. (—A— has a functionality of two).

If a small amount of a monomer with a functionality of *three or more* is included branching can occur:

$$-A-A-A-A-A-B\begin{smallmatrix} A-A-A- \\ \\ A-A- \end{smallmatrix} \qquad (-B \text{ has a functionality of three})$$

If a large amount of trifunctional monomer is present, a network in three dimensions is formed (Fig. 12).

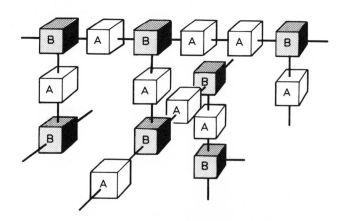

**Fig. 12** A cross-linked polymer network.

The polymer is then said to be *cross-linked*.

The effect of functionality may be summarized thus:

| Functionality of monomer | Type of product |
| --- | --- |
| ONE | Simple compound |
| TWO | Linear polymer |
| THREE (*minor ingredient*, when principal monomer is difunctional) | Branched polymer |
| THREE (*major ingredient*) | Cross-linked polymer |

The importance of functionality lies in the different properties that may be expected from the types of polymer produced. Let us consider a piece of solid polymer again.

If the polymer is linear, the long chains will be randomly packed, some coiled, some extended and several entangled with one another. Remember that, since rotation can occur about any single bond, a polymer chain is extremely flexible. The atoms will not all lie in a straight line and the chain may be fully extended (Fig. 7a), partially coiled, fully coiled (Fig. 7b) or even, presumably, knotted! However, heat sets the molecules moving and small liquid molecules may penetrate between the chains. Both help to sort out the tangle, i.e. the polymer will melt or dissolve. The same is true for lightly branched polymers of low molecular weight.

In a cross-linked polymer, every chain is probably connected to at least three others, so that the solid lump is one molecule. To separate the chains we have to *break* several covalent bonds. Thus a cross-linked polymer is at first softened by heat, as the segments between cross-links get a limited freedom of movement, but the next stage is not melting: it is decomposition. Again, while liquid molecules can penetrate into the holes in the cage-like structure, causing it to swell, they cannot dissolve the polymer. Since dissolving implies separating the molecules from one another, this is clearly impossible. There is only one molecule.

Linear polymers, because of their ability to melt and solidify as many times as may be required, are spoken of as *thermoplastic* polymers. Linear polymers which subsequently cross-link on heating are described as *thermosetting* polymers.

## Crystallinity in polymers

Up to this point, no firm statement has been made about the nature of linear synthetic polymers as solids. Their resemblance to the amorphous natural resins has been mentioned and, in the previous section, reference to random packing has inferred an amorphous nature. At the same time, the first synthetic linear polymer described, polyethylene, was said to be 'markedly crystalline'.

Since crystallinity requires the arrangement of atoms and molecules in a three-dimensional ordered pattern, and since polymers contain large, complex, flexible molecules, it may be difficult at first to see how polymers can be crystalline. Crystallinity in a polymer involves the parallel arrangement of lengthy segments of uncoiled uniform polymer chains on a three-dimensional scale. It does not necessarily involve complete chain lengths. A crystalline area in a polymer can be likened to a bundle of rods. The conditions for such an alignment to occur in a polymer are these:

(1) The lengths of chain concerned should consist of regular repeating units arranged head-to-tail and should contain no bulky side-groups, which could prevent the close parallel packing of the extended chains.
(2) There should be sufficient intermolecular forces of attraction – resulting from polarity – to hold the chains together in the crystalline pattern, when the chains align as closely as their side-groups will let them.

These conditions are quite exacting, so that it is not surprising that even polymers whose molecular structures appear to be ideal for crystallinity, should show substantial areas of non-crystallinity in the solid form. These areas are due to the presence of coiled and tangled molecules, which could not uncoil in response to intermolecular forces during the solidification process. Equally, less suitable polymers, forming largely amorphous solids, can show areas of crystallinity where, during the solidification process, chance has decreed that chain segments of the right type should come together. Thus no polymer is wholly crystalline and few are completely amorphous.

Marked crystallinity introduces a number of objectionable features into a polymer, from a paint point of view:

(1) Lack of transparency, due to refractive index differences between regions of the polymer solid (see Chapter 6).
(2) Sharp, and often high, softening temperatures. Gradual softening over a range of temperature is advantageous in stoving paints. It will permit minor irregularities to flow out, without causing 'sags'. The glass temperature of a polymer used in an emulsion paint should be lower than the paint's drying temperature (see Chapter 11).
(3) Insolubility in all but the most polar liquids, which have sufficient attraction for the polymer molecules to separate them. In some cases, not even these are effective.

Thus for paint purposes, largely amorphous polymers are required. The tendency for a polymer to crystallize can be reduced and the softening temperature lowered by:

(1) making the arrangement of repeating units an irregular one, e.g. by copolymerization,

(2) introducing bulky side groups,
(3) spacing out the polar groups that provide the strong intermolecular forces,
(4) introducing a plasticizing liquid of high boiling point, which reduces inter-molecular attractions (being itself attracted by these forces) and increases the spacing between polymer chains.

Some polymer structures will now be considered for their effect on crystallinity and general properties.

*POLYACRYLONITRILE* has the ideal structure for crystallinity: uniform chain; relatively small, highly polar side groups. The inter-molecular attractions are reinforced by hydrogen bonding:

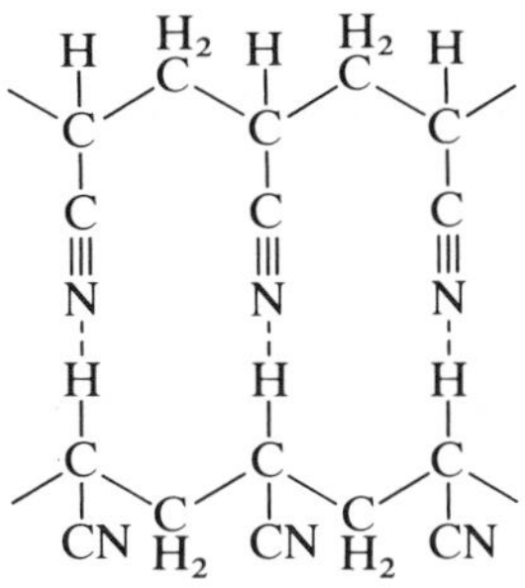

This polymer has high softening temperatures and is insoluble in all but a handful of highly polar liquids with marked hydrogen bonding capacity (e.g. $H \cdot CO \cdot N(CH_3)_2$, dimethyl formamide). Both properties are the direct result of the extreme difficulty of separating the chains and breaking the crystal structure. The polymer is unsuitable for paints, but has ideal properties for a synthetic fibre ('Acrilan').

*POLYETHYLENE* is a non-polar polymer with weak inter-molecular forces, but the simplicity of its structure

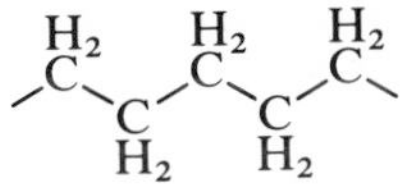

allows very close packing of the chains. The polymer is therefore markedly crystalline but, since relatively little energy is required to separate the molecules, it has a low softening temperature (105–115 °C). It is completely insoluble at room temperature, though swollen by hydrocarbons. It is therefore unsuitable for paints and fibres, but is a useful plastic.

*POLYVINYL CHLORIDE*

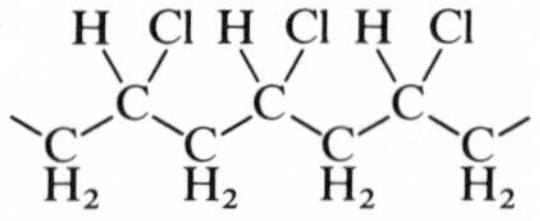

This polymer is similar to acrylonitrile, in that it contains small polar side groups on a regular chain. It is therefore markedly crystalline, with a high softening temperature and solubility in few solvents. It has, however, found limited use in paints in two ways.

In the first, the crystalline polymer is used in particulate form and the particles are dispersed in a liquid–solvent or plasticizer–capable of penetrating the particles at higher temperatures (*c.* 170 °C) and integrating them into a film. Crystallinity in the film is reduced and flexibility increased by the plasticizer, or by plasticizing resin dissolved in the solvent. Such mixtures are called *organosols* or *plastisols* (see Chapter 11).

The second method is to copolymerize vinyl chloride with about 10–20 per cent vinyl acetate. The copolymers are soluble in ketones and esters.

*POLYVINYL ACETATE*

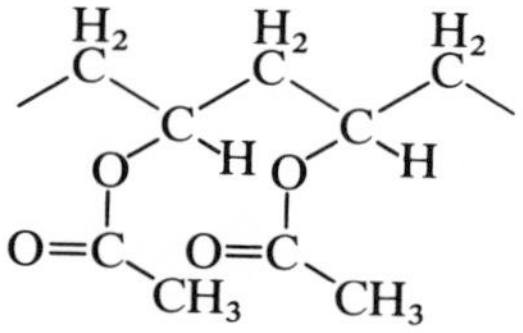

Although the side group of this polymer has some polarity, the bulk and flexibility of the group lead to irregularity and wide spacing of chains. The polymer is amorphous, readily dissolved and highly suitable for paints. Softening temperature varies from about 70 °C, for low molecular weight samples, to 125 °C for higher molecular weight materials. This variation with molecular weight in ease of softening (and ease of solution) is common to all linear amorphous polymers. Higher molecular weights mean more polar groups per molecule, a greater number of points of attraction between molecules and an increased opportunity for tangling. Thus more needs to be done to separate the molecules.

*POLYSTYRENE.* This polymer also has a bulky side group, but the side-chain is of relatively low polarity:

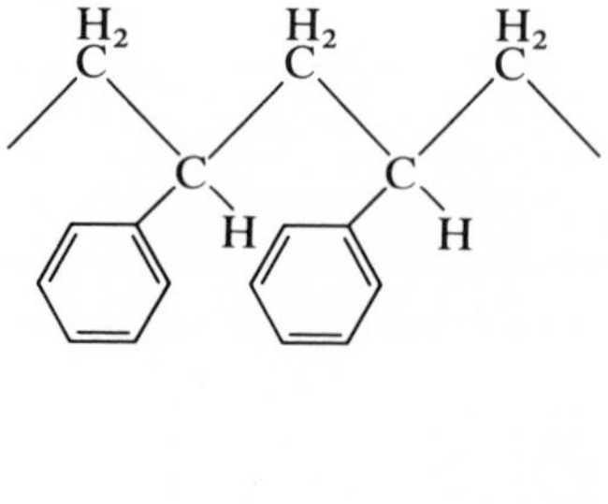

Both factors are against crystallinity and the polymer is indeed amorphous and soluble. In this polymer, however, the side groups are rigid and they restrict considerably the bending and coiling of the polymer chains. Except at high molecular weights (unsuitable for paints), polystyrene is therefore weak and brittle and styrene copolymers (sometimes styrene is copolymerized with drying oils and alkyds) are more suitable as paint resins.

This illustrates a general principle: rigid groups – in the polymer backbone, or attached to it – make a polymer inflexible.

*Condensation polymers* are governed by the same rules. Chains are usually less regular, but more polar. At least two polymers show such high crystallinity that they have been used for new fibres: 'Nylon' (polyamide) and 'Terylene' (polyester).

# Six

# Colour

## Light

Light is a form of energy. Our chief source of light is the sun, a vast ball of matter, kept at a fierce heat by the energy produced in a continuous series of chemical and nuclear reactions proceeding in it. The energy released in these reactions is also felt by us as heat. Heat and light are produced in the same process; they are different manifestations of energy.

In fact they form part of a large series of radiations, all different forms of energy. Energy is radiated in waves or pulses, alternating between maximum and zero intensity – the crests and troughs of the waves. These radiations all move at the same high speed: approximately 186 000 miles per second, through air or vacuum. They differ, however, in *frequency*. The frequency is the *number* of waves that move past a fixed point in one second. Since all the radiations move at the same speed, the more waves that move past the point per second, the shorter they must be, i.e. the *wavelengths* must vary. The wavelength is the distance from crest to crest or trough to trough. These facts are summarized in Fig. 13.

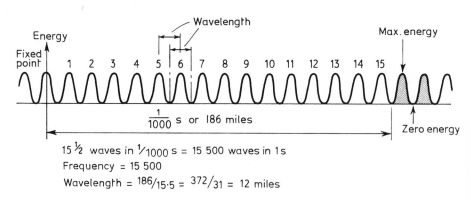

15 ½ waves in ¹⁄₁₀₀₀ s = 15 500 waves in 1s
Frequency = 15 500
Wavelength = ¹⁸⁶⁄₁₅.₅ = ³⁷²⁄₃₁ = 12 miles

**Fig. 13** Electromagnetic radiation: energy, frequency and wavelength.

The radiations referred to above are known as *electromagnetic radiations*.

Their wavelengths vary as follows:

| | |
|---|---|
| Less than 0·1 nm | Gamma rays |
| 0·001–30 nm | X-rays |
| 30–400 nm | Ultra-violet light |
| 400–700 nm | Visible light |
| 700 nm–0·5 mm | Infra-red rays (radiant heat) |
| 0·1 mm–10 m | Radar and television |
| 0·1 m–20 km | Radio waves |
| Above 20 km | Long electrical oscillations |

It is difficult to imagine all these phenomena as being of the same type and differing only in wavelength and frequency, because they are recognized in completely different ways. Visible light seems different, because it is the only part of the series which affects the eyes in such a way that they transmit signals to the brain, which the brain in turn interprets as a picture.

## Reflection

That light travels in straight lines is clear, from the fact that a beam can be produced if light is passed through a slit or hole, and from the well-known explanation of eclipses. If a beam of light shines onto a surface, it will be wholly or partly reflected. Light meeting a smooth and level surface at a certain angle, will leave it at a similar angle. This is usually expressed by saying that the *angle of incidence* is equal to the *angle of reflection* (see Fig. 14).

By 'surface', we mean a boundary between two different types of substance. For example, light will be reflected from the bottom face, as well as the top face of a plate of glass, since both are glass/air boundaries (Fig. 15).

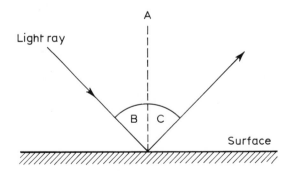

**Fig. 14** Reflection: angles of incidence and reflection.
    A. The normal (at right angles to the surface).
    B. Angle of incidence.
    C. Angle of reflection.

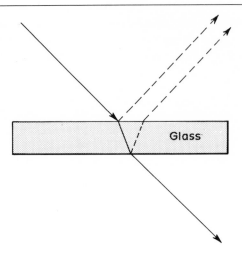

**Fig. 15** Reflection at glass-air interfaces.

One of the consequences of this law of reflection is that it explains the 'gloss' on a paint film. A perfectly smooth paint film is like the surface in Fig. 14: light from an object is reflected uniformly to the eye of the observer. Since the eyes receive a complete picture of the object, the observer sees it in sharp outline and the film looks glossy. If the surface is not smooth (Fig. 16), parts of the object are seen clearly by reflection, but parts are 'lost' since the light is reflected elsewhere and not to the eye. The overall result is a blurred picture and the observer decides that the gloss is low.

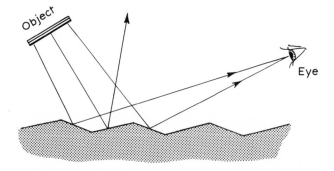

**Fig. 16** Reflection at a surface of low gloss.

### Refraction

When light meets a surface, some will be reflected, but the remainder will pass below the surface where, either it is absorbed and the object is said to be opaque, or it passes through the object, which is said to be transparent or translucent. Now the speed of light varies with the medium through which it is passing and its speed through a transparent object will certainly be slower than its speed in air. The effect of this is that the path of the light is 'bent' at each boundary between the object and the air (Fig. 15). The 'bending' process is called *refraction*. For example, an object at the bottom of a bowl of water always appears closer than it is. In Fig. 17, light passes from object A at the bottom of the water to the surface at C, where it is refracted and then enters the eye. The eye and brain assume that the light has travelled in a straight line and therefore 'see' the object at B.

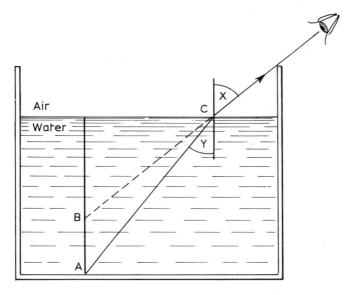

**Fig. 17** Refraction in water.

The amount of refraction at a boundary is expressed by the *refractive index*:

$$\text{refractive index} = \frac{\text{sine } \angle X}{\text{sine } \angle Y} \quad (\text{Fig. 17}) = \frac{\text{speed of light in air}}{\text{speed of light in other medium}}$$

The refractive index of water is 1·33. For diamond it is 2·42, but rutile titanium dioxide has a higher refractive index than any other transparent material: 2·76.

Most surfaces involved in reflection or refraction are not flat. The behaviour of light at a curved boundary can be assessed by drawing a tangent to the surface, at the point at which the light strikes it. If it is now imagined that this tangent (or tangential plane) is the surface, the imaginary surface is thus a flat one and all the laws of reflection and refraction for level surfaces can be applied (Fig. 18).

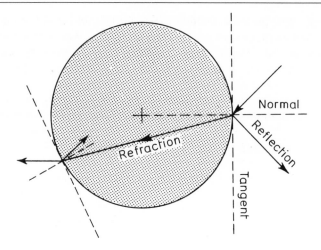

**Fig. 18** Reflection and refraction at curved surfaces.

## Diffraction

You may have observed that if water waves meet an obstacle of some kind, the waves that pass by the obstacle have the ability to pass round it to some extent, disturbing the still water in the lee of the object. In other words, waves can get round corners. Sound waves can get round the corners of buildings in city streets. The phenomenon is called *diffraction*.

The extent of diffraction depends on the wavelength. In considering the reception of radio waves, the position is slightly complicated by the possibility of reflecting these waves off the ionized Heaviside and Appleton layers, or even off satellites. In the daytime, reflection of wavelengths below 10 000 metres is inefficient and the 'groundwaves' (those travelling from place to place by the most direct route) are more important. Foreign stations are easily heard on medium and long wavebands (200–2000 m), though one might expect the earth's curvature, hills, etc., to prevent the passage of the waves. Clearly diffraction occurs. Short wave (10–100 m) radio communication is limited to shorter distances. Television signals (about 5 m) are affected by hills, though sometimes weak reception is possible on the 'shadow' side of the hill. With UHF television signals, such as BBC 2 (about 0·5 m), hills are more effective in completely preventing reception. Thus diffraction is not important when the size of the obstacle is much greater than the wavelength and is most important when the size of the obstacle is similar to the wavelength.

Light has such short wavelengths (less than 1 $\mu$m), that diffraction is not normally noticed. It is, however, impossible to get a shadow with absolutely sharp edges, even with a point source of light. Light, therefore, can pass round obstacles: it can suffer diffraction. A phenomenon produced by the diffraction of light, is the halo sometimes seen *closely* surrounding the sun or moon. This is due to diffraction of light by small droplets of water in thin layers of cloud.

These facts are very relevant to paint films. They can explain, for example, why a mixture of two transparent resin solutions can dry to give a milky film, if the resins are *incompatible*. Resins that are incompatible are simply resins that will not mix. The solvents help them to mix in solution, but as the liquids evaporate, the molecules of the two types of resin tend to congregate together in separate groups. Complete separation into two layers does not occur, because the viscosity is, by this time, too high. Instead, we have a dispersion of droplets of one resin inside a film of the other.

In this resin film are hundreds of interfaces or boundaries between the two resins. If the resins differ in refractive index, refraction will occur at every interface. Fig. 19 shows how light can be returned from a film by successive refractions. Reflection will also occur and by this means some of the light will be returned towards the surface of the film. It is often forgotten that diffraction will also occur at the edges of the droplets. The net effect of multiple diffractions is extremely complex: it is sufficient to say that diffraction can play an important part in turning the light back towards the film surface.

The overall result is that white light is returned from the film to the eyes of the observer, who concludes that the film is white. The greater the difference in refractive indices and the larger the number of particles up to a limiting concentration, the more milky and less transparent the film will appear. Transparent materials can appear white in powder form, because white light is reflected, refracted and diffracted at the many solid-air interfaces.

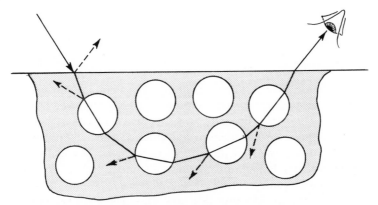

**Fig. 19** Light returned from a 'white' film.

## Colour

If two electromagnetic radiations are very different in wavelength, the radiations will be different in type. If two rays of visible light differ slightly in wavelength, they will have different effects on the eye, i.e. they will appear to be different in *colour*. There is a *gradual* change in colour from violet at about 400 nm, through blue, green, yellow and orange to red at about 620 nm. Light at any unit wavelength is almost imperceptibly different from light at the next unit wavelength and one colour merges into the next, so that the number of different colours cannot be counted. All these wavelengths are present in sunlight as a mixture and the eye registers this mixture as the colour we call white.

Since the refractive index of a transparent substance varies with the wavelength of light, it is possible to separate the wavelengths. When light passes into glass, each wavelength is bent by a different amount. As the light rays pass out of a glass plate, they are bent back and converge (Fig. 15), but if the two sides of the glass are not parallel, as in a prism, the rays diverge (Fig. 20). The range of colour produced is called a *spectrum*.

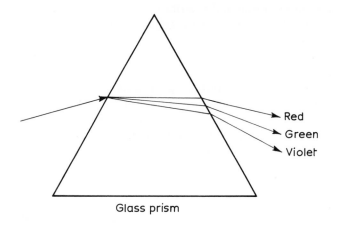

**Fig. 20** Prism and spectrum.

When a coloured object is illuminated by white light, it selectively absorbs some of the wavelengths and transmits others. The eye receives the transmitted wavelengths and summarizes them as one colour. This colour is dominated by the main wavelengths transmitted, which determine its basic hue. This hue is modified by undertones of the other wavelengths, which make it an individual shade. Thus there are many red shades; some are bright, some dark; some have a yellow undertone (e.g. scarlet) and some a blue one (e.g. crimson). A truly black object absorbs all the wavelengths and a truly white one, none of them.

## Colour chemistry

It is interesting to see if there is any connexion between the chemical structures and the colours of compounds. Among organic compounds, we find that coloured chemicals usually contain certain groups which are called *chromophores* (colour-bearers). These are:

−N=N−

azo group

$-N\overset{\displaystyle O}{\underset{\displaystyle O}{\diagup}}$

nitro group

−CH=N−

azomethine group

$\diagdown$C=O

carbonyl group

−N=N−
‖
O

azoxy group

$\diagdown$C=S

thio group

−N=O

nitroso group

$\diagdown$C=C$\diagup$

ethenyl group

They are usually found attached to aromatic rings, such as benzene, or one of the more complex ring structures, such as

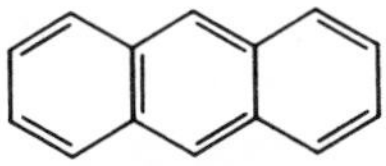

anthracene

Coloured molecules have their colour intensified or modified by certain other groups, known as *auxochromes*:

$-N\overset{\displaystyle R}{\underset{\displaystyle R}{\diagup}}$ $\qquad$ $-N\overset{\displaystyle H}{\underset{\displaystyle R}{\diagup}}$ $\qquad$ −NH$_2$ $\qquad$ −OH $\qquad$ −OCH$_3$

| tertiary amino group | secondary amino group | primary amino group | hydroxyl group | methoxy group |

−I $\qquad$ −Br $\qquad$ −Cl

iodo $\qquad$ bromo $\qquad$ chloro groups

Thus the dye alizarin (red) contains two benzene rings, two carbonyl groups (chromophores) and two hydroxyl groups (auxochromes):

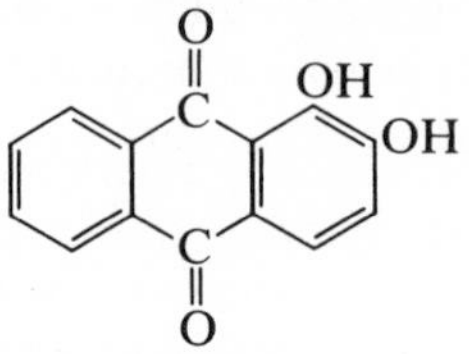

alizarin

It should be noted that all the groups that influence colour contain electronegative atoms, or double bonds or aromatic rings. All of these attract higher than average densities of electrons, indicating that colour is associated with compounds which have high localized electron distributions. The energy from electromagnetic radiations is, in fact, absorbed by the electrons of a molecule. Colourless compounds absorb only in regions outside the visible spectrum.

Colour in inorganic compounds is again connected with the behaviour of electrons. The elements of Groups IIIA to IB in the Periodic Table, known as the *transition elements*, contain special electrons which, under the influence of ligands, can absorb energy in the visible region of the spectrum. Ligands are attached groups, ions or molecules, such as the sulphate ion and water molecules in hydrated copper sulphate, $CuSO_4 \cdot 5H_2O$. The exact region in which the absorption takes place is determined by the nature and positioning of the ligands. Thus $CuSO_4 \cdot 5H_2O$ is pale blue, but $CuCl_2 \cdot 2H_2O$ is green.

In other inorganic compounds, the amount of energy that can be absorbed is sometimes sufficient to promote the transfer of a valency electron from one atom or ion to another, e.g. absorption causing

$$Pb^{2+}I_2^- \longrightarrow Pb^+I^- + I$$

is responsible for the golden colour of lead iodide, $PbI_2$. This type of *charge transfer* is responsible for the intense colour of some inorganic compounds, especially when no transition element is present, and it occurs particularly with oxides, sulphides, iodides and bromides.

## Colour mixture

Sometimes the eye observes a colour which is not to be found anywhere in the spectrum, e.g. grey, brown or purple. Since any colour is the sensation stimulated by a mixture of wavelengths, it will be worth while to see what rules govern such mixtures and how new shades are produced. The first point to note is that very different results are obtained by mixing coloured lights on the one hand and coloured pigments on the other. Let us examine these two types of mixing in more detail.

### Additive mixing

Suppose we take a white screen into a darkened room and shine coloured lights on to it. Since the material is white in daylight, it therefore reflects all wavelengths. Thus a yellow beam will give a yellow spot on the screen and a blue-violet beam, shining on another part of the screen, will give a blue-violet spot. If both lights are then focused on to the same part of the screen, we see *white*.

The yellow and blue-violet lights are *complementary* to one another and their mixture in the right proportions is registered by the eye as white. Other complementary pairs can be seen by drawing straight lines through the centre spot of Fig. 21 (e.g. red and peacock blue).

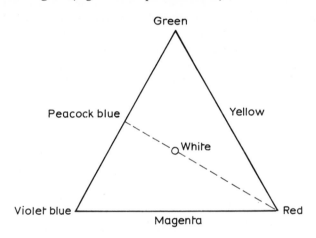

**Fig. 21** Additive colour mixing diagram.

Red, blue-violet and green are called the three *primary* colours. Secondary colours are produced along the sides of the triangle, by mixing any two primaries in different proportions, e.g. mixtures of blue-violet and red light, vary from purple through magenta, to rose-pink, as the proportion of red increases. These colours do not occur in the spectrum. Other colours can be made by mixing across the triangle.

Remember that light is energy and the white screen reflects all the light. Thus if two lights of different colour shine on to the same part of the screen, both are reflected and the *amount* of energy reaching the eye is more than would reach it from one light only. Consequently the colour seems *brighter*. Because the lights are added together to give more energy and brighter colour, this type of colour mixing is called *additive mixing*.

## Subtractive mixing

If we mix a blue pigment and a yellow pigment we do not get white, we get *green*. This is because both pigments *absorb* light. The blue pigment absorbs practically all of the orange and most of the red and yellow wavelengths from the white light falling on it. The yellow pigment absorbs almost all the violet, the remainder of the red and nearly all the blue. Looking at the list of the six principal colours of the spectrum, given in the section headed **Colour**, we see that green is the only colour not absorbed by either pigment.

Both pigments are extremely fine in particle size and the particles of both are uniformly distributed over the area covered by the mixture. Light falling on the area therefore passes through a double 'filter' and the only wavelengths that can be returned to the eye are those that are absorbed by neither pigment. Note that if both pigments are dull (i.e. absorb strongly), the green will be even duller, because less light energy will reach the eye.

Mixing pigments, blue and yellow are not complementary, but there are complementary colours, shown opposite one another in Fig. 22, e.g. red and green. Mixtures of complementary colours in the right proportions do not produce white; instead, dark *grey*. They cannot produce white because they do not reflect *enough* light energy. White and black are two extremes: complete reflection and none, maximum light energy and nil. In between come all the shades of grey, getting darker and darker as less light is reflected. A neutral grey reflects all the wavelengths of white light, but it does not reflect *enough* light to give the full white colour.

**Fig. 22** Subtractive colour mixing diagram.

In the pigment circle (Fig. 22), a number of pigments are listed. The pigment colours do not correspond to pure colours, but lie between them. If two pigments that are not complementary are mixed, the colour produced will lie between them, on the shorter arc of circle joining them, e.g. red and blue can give violet. If the two pigments are joined by a straight line, the closer that line lies to the centre of the circle, the darker the colour will be.

Because each individual pigment in a mixture subtracts some of the wavelengths and some of the light energy from the light falling on the mixture, this type of colour mixing is called *subtractive mixing.** It is the type of mixing that the paint maker must carry out to make a paint with a given colour (Chapter 8).

* Strictly speaking, subtractive mixing occurs only with dyes, which absorb light, but cannot reflect it. Pigment particles reflect some light from their surfaces.

# PART TWO

# Applied science

# Seven

# Paint: first principles

## Wet paint: paint in the can

All objects are vulnerable at their surfaces. It is the surface of any article that makes continual contact with the corroding (or oxidizing) air. The surfaces of objects left in the open bear the brunt of the sun, rain, fog, dew, ice and snow. Under these conditions iron rusts, wood rots (or shrinks and cracks) and road surfaces crack and disintegrate. These, and more sheltered objects, suffer the wear of daily use, scratches, dents and abrasions – at their surfaces. To prevent or to minimize damage, man applies to these surfaces various coatings designed to protect them. Coatings can also be used to decorate the articles, to add colour and lustre and to smooth out any roughness or irregularities caused by the manufacturing process. Thus the function of any surface coating is twofold: to protect and to decorate.

There are many surface coatings that do this: wallpaper, plastic sheet, chrome and silver plating. No coating material is more versatile than paint, which can be applied to any surface, however awkward its shape or size, by one process or another. Paint is a loosely used word covering a whole variety of materials: enamels, lacquers, varnishes, undercoats, surfacers, primers, sealers, fillers, stoppers and many others. It is essential to grasp at once that these and other less less obviously related products, such as plasters, concrete, tars and adhesives, are all formulated on the same basic principles and contain some or all of three main ingredients.

First a *pigment* may be included. Pigments have both decorative and protective properties. The simplest form of paint is whitewash and, when dry, whitewash is nothing more than a pigment – whiting (calcium carbonate) – spread over a surface. It decorates and to some extent it protects, but it rubs off. So most paints contain the second ingredient, a resin polymer, *film-former* or *binder*, to bind together the pigment particles and hold them on to the surface. If the pigment is left out, the film-former covers and protects the surface, decorating it by giving it gloss or 'sheen'. It is difficult to attach coatings that are not fluid to any but the simplest of surfaces: those that are flat or gently curving. The fluidity of paint permits penetration into the most intricate crevices. It is achieved by dissolving the film-former in a solvent, or by colloidal suspension of both pigment and film-former in a diluent. Thus the third basic ingredient of paint is a *liquid*. Often the film-former/liquid mixture is called the *vehicle* for the pigment.

If the pigment is omitted, the material is usually called a *varnish*. The pigmented varnish – the *paint* – is sometimes called an *enamel, lacquer, finish* or *topcoat*, meaning that it is the last coat to be applied and the one seen when the coated object is examined. *Lacquers* are normally thermoplastic solution paints or varnishes, but the term is sometimes (confusingly) used to describe all clear woodfinishes. *Enamels* are normally thermosetting paints, hard, with a superficial resemblance to vitreous enamels.

Paint applied before the topcoat is called an *undercoat*. Some undercoats may be briefly defined as follows:

(1) *Fillers or stoppers* are materials of high solid content, used to fill holes and deeper irregularities and to provide a level surface for the next coat.
(2) *Primers* are applied to the filled or unfilled surface, to promote adhesion, to prevent absorption of later coats by porous surfaces and to give corrosion resistance over metals. Special pigments improve the anti-corrosive properties.
(3) *Surfacers* (called 'undercoats' in the house painting trade) are highly pigmented materials containing large quantities of extender (see below). They are easily rubbed smooth with abrasive paper. They provide the body of the paint film, level out minor irregularities in the substrate and must stick well to primer and topcoat.
(4) *Primer-surfacers* are surfacers that can be applied direct to the object's surface (the substrate).
(5) *Sealers* are clear or pigmented materials applied in thin coats to prevent the passage of substances from one coat of paint to another or from the substrate into later coats. They can be required to improve adhesion between coats, where this is otherwise weak.

All these materials are formulated on the principles described above. These are illustrated in the diagram opposite, which also lists some of the minor ingredients of a paint. Some of the terms in the diagram need a little further explanation at this stage:

*Pigment:* Any fine solid particles that do not dissolve in the varnish. If they do not provide colour they are called *extender* particles. Extenders are much cheaper than prime pigments and can carry out many useful functions, e.g. improvement of adhesion, ease of sanding and film strength.

*Film-former:* When the coating is dry the film-former is a polymer, but in the wet sample it may be merely the chemical ingredients which react to form the final dry polymer.

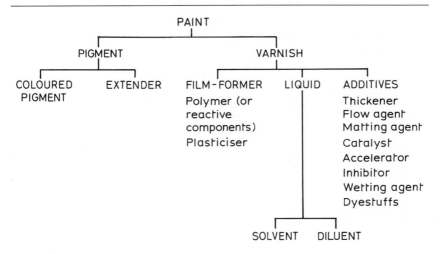

*Liquid:* Some of the liquids of the paint are often withheld from the paint container and supplied separately as a *thinner*. The user adds thinner to adjust the consistency to his requirements.

*Additives:* Small quantities of substances added to carry out special jobs, such as the improvement of surface appearance.

## Application

These are the basic ingredients, but much care in formulation will be required to produce a paint that will be easy to apply under changing conditions and pleasant to look at when dry. It is not within the aims of this book to discuss the methods of application that are available. They are well described elsewhere (see Appendix). Suffice it to say that the paint may be put on by brushing, by roller, by a whole variety of methods of spraying (compressed air, airless, electrostatic and aerosol spraying) by dipping, electro-deposition, curtain-coating and flow-coating. Either the article is immersed in the paint, the excess of which is then allowed to drain off, or the correct quantity of paint is applied to the article and must not drain, 'run' or 'sag'.

In all cases, any irregularities in the wet film caused by the method of application must flow out to leave a smooth surface.

The first problem is to make the paint easy to handle while it is being applied. Different application methods require paints of different consistency, but in all cases the principle is the same: the greater the content of dissolved polymer in a paint, the more viscous it will be. The weight percentage of involatile material found in the paint is known as the solids content or 'solids' of the paint. Either by adjusting the 'solids' of the paint, or by altering the balance of the types of liquid used, the formulator brings the consistency or viscosity of the paint to that required.

So far so good.

The difficulty lies in the next stage. Most methods of application leave some irregularities in the wet film surface: brush marks, spray mottle, roller stipple, and so on. These are the methods that apply the correct amount of paint, which must flow at first to remove the irregularities, and then stop flowing, to prevent 'running' of paint on vertical surfaces. This change in the rate of flow is usually brought about by evaporation of solvent, causing a rise in 'solids' and hence a thickening of the consistency. Flow dwindles, until it is scarcely occurring at all. Then the drying mechanism takes over to set the film. A very careful balance of solvents is required to do this satisfactorily and more will be said about this in Chapter 9.

An alternative method, often used in conjunction with solvent evaporation, is to include some material in the paint which gives it abnormal viscosity characteristics, so that it is fluid while being agitated, but thickens up over a period when the agitation stops. This type of paint is said to be 'thixotropic'. More detail is given in Chapter 10.

If the article is allowed to drain after being coated, as in dipping, then it must drain evenly and not so rapidly that the film thickness becomes too low. If possible, thicknesses at the top and bottom of the article must vary only slightly, in spite of the downward drain of paint. Again the same principles must be used to slacken the flow.

With spray application in particular, the position is further complicated by the fact that the paint reaching the surface does not have the same composition as that leaving the spray-gun. The paint is broken up into thousands of fine droplets as it leaves the gun, each droplet presenting a surface – at which evaporation occurs – that is large compared with the droplet's volume. A great deal of liquid can be lost and this must be taken into account in formulating the paint.

It is necessary to point out that this loss of solvent – all of which has been paid for – is only accepted by the paint user on sufferance, as an inevitable consequence of the process. He is constantly looking for new materials which will contain less wasted material: in other words he wants paints with higher 'solids'. Increasingly legislation is being enacted to reduce the amount of organic material which can be exhausted into the atmosphere, thereby polluting it. Furthermore, solvents come from petroleum, the world stocks of which are limited and must be conserved. For all these reasons, the paint formulator chooses his ingredients to give the highest 'solids' possible. To keep the paint manageable at high solids, the formulator has to keep the polymer molecular weight down, since, as will be seen in Chapter 9, this reduces the viscosity. Alternatively (Chapter 9 again), he must take the polymer out of solution (if this is practicable and suitable) and make an 'emulsion' paint.

## Dry film properties

An outline paint formula has been described and it has been shown that application presents problems which influence the formula. After application comes drying, but before this is considered, the properties required from the dry film must be discussed, since these influence the choice of method of drying.

If the dry paint film is to be a useful one, it must stick to the surface beneath, be hard enough and flexible enough for the purpose for which it is used and must retain most of its protective and decorative properties for a long period. The paint should be capable of repair or renovation. Let us now take these requirements separately and see how they are achieved.

### Adhesion

We have seen that most molecules or atoms have some attraction for one another. The strength of this attraction varies a lot with the atoms concerned, but all inter-molecular attractions have one thing in common: they operate over comparatively short distances (less than 10 nm) and become weaker the farther apart the atoms are within that short range. At distances above 1 nm their contribution to adhesion is negligible.

Since these attractive forces are the ones that make things stick together, the first problem is to get the molecules within range, i.e. the paint film must 'wet' the surface, displacing air and all the other adsorbed materials. The *critical surface tension* ($\gamma_c$) of a smooth solid surface is a numerical measure of the ease or difficulty of wetting it. It is equal to the highest surface tension possessed by any liquid that will spread spontaneously when placed on it. If a paint is to wet the surface, it must have a surface tension *equal to or lower than* the critical surface tension of the solid. Plastics can have very low critical surface tensions (e.g. polytetrafluoroethylene, PTFE = 18.5; polyethylene = 31 dynes/cm), which limit considerably the choice of solvents for paints to be applied to them.

Clean metals generally have higher critical surface tensions (>73 dynes/cm), but even the thinnest possible layer of adsorbed oil or grease can dominate the surface (e.g. $\gamma_c$ for 'clean' tinplate cans has been measured as 31 dynes/cm) leading to wetting problems. Even if the paint wets the contaminated surface, loosely adhering grease, dirt or rust is a menace since, although the paint may stick well to the adsorbed layer, the latter can easily come away from the surface taking the paint film with it. For this reason, the cleaning of surfaces before painting is to be recommended. The practice of rubbing undercoats with abrasive paper, not only provides a level surface for the next coat, but also cleans the surface, removing materials difficult to wet or poor in cohesion.

Semi-porous undercoats or surfaces can offer good adhesion by a different

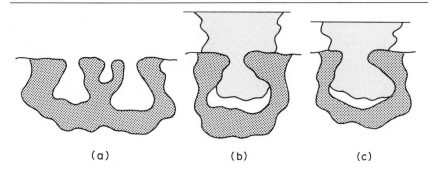

(a)                    (b)                    (c)

**Fig. 23** Mechanical adhesion.
A. Cross-section of porous surface.
B. Wet paint on surface.
C. Some contraction of paint film after drying.

mechanism. Providing the liquid paint can displace air from the crevices, the change from liquid to solid will mechanically lock or 'key' the paint on to the surface (Fig. 23). If the clean surface is not porous, but has been wetted well, adhesion will depend on the strength of the paint-substrate inter-molecular attractions. Sometimes these can be considerably enhanced by the inclusion of particular chemical groups in the molecules of the paint film-former, e.g. carboxyl groups promote adhesion to metals.

### Hardness, toughness and durability

These properties are connected in that cross-linked films are generally harder, tougher and more durable than those that are not. High molecular weight linear polymers, however, produce harder, tougher, more durable films than their low molecular weight equivalents. If the molecular weight is high enough, a polymer will be as hard, as tough and as durable in linear form, as it can be in cross-linked form.

### Flexibility

Hardness often goes with brittleness, and brittle films – particularly those subjected to temperature changes – do not survive long, failing by cracking and then flaking off if adhesion is only mediocre.

Flexibility is introduced into cross-linked films by spacing out the cross-links, so that the whole structure becomes a looser, more open cagework of molecules.

Linear polymers are made more flexible by admixture with smaller molecules, which separate the large polymer chains, reduce attractive forces between them and act as a lubricant, allowing the polymer molecules to slide past one another more easily. These smaller molecules are called 'plasticizers'.

Some monomers give more flexible polymers than others. They may be co-polymerized into brittle polymers to increase flexibility.

### Loss of decorative properties due to weathering

Gloss may go, the surface ultimately becoming powdery, or the colour may fade or darken. Ultra-violet rays and water, as well as oxygen in the air, gradually attack paint films, breaking chemical bonds and causing polymer molecules to break up into smaller fragments. The surface gradually wears away, roughens and the gloss disappears. Polishing will often rub it smooth again. This type of wear is reduced by choosing resistant polymers and by keeping the proportion of pigments down. If the pigment volume concentration in the film is high, it will only require the breakdown of a small amount of polymer to cause the detachment of surface pigment particles. The powdery surface is said to have 'chalked'.

Colour fastness is dependent on the use of high quality pigments though complete satisfaction is not guaranteed unless the pigments have been tested in the binder that is to be used in the proposed paint.

### Ease of repair and surface renovation; solvent resistance

Paint users often require to be able to make good damage to paint films. This 'damage' may be no more than excessive dirt or dust pick-up during application, spoiling the appearance. It might be shallow scratching. In either case, all that may be needed is light abrasion followed by polishing. To polish well, a surface should be capable of being softened by the heat generated, so that it will partially flow, evening out irregularities. If the polymer is cross-linked, its rigid network of chains of atoms restricts this flow. The best polishing properties are usually shown by linear polymers. Their ability to reflow is demonstrated in the acrylic lacquer process for cars, in which the lacquer surface is rubbed with very fine emery paper to remove dust and spray mottle. Subjecting the lacquer to heat in an oven reflows the scratches, so that the surface is bright and glossy again.

Linear polymers are also soluble in solvents: cross-linked polymers are not. Thus the surface of a linear polymer film can be levelled by light rubbing with a pad soaked in a solvent mixture that will *just* dissolve the polymer. This process is called 'pulling over' and is used in the furniture industry.

If the paint film is badly damaged, it must be patched. A linear polymer film can be patched by spraying in the damaged area with the same paint. The solvents redissolve the edges of the existing film, marrying the new area into the old. Cross-linked polymer films will not redissolve and so will not patch. The complete area must be rubbed down and repainted. Any attempted patch will be at best partially successful, with all sorts of trouble likely to occur at the join between patch and original film.

It follows that linear polymer films will have poor resistance to solvents (present in many household articles, e.g. nail varnish). Cross-linked films will be more resistant.

If a paint is to be formulated for a particular purpose, the properties required from the dry film are defined first. These properties and the desired minimum solids of the paint, may well determine whether the dry film is to contain a linear or a cross-linked polymer. When the type of polymer film has been selected, the choice of drying mechanism will have been restricted to a narrow area.

## Dry paint: how paints dry

In the can of paint the mixture of substances must remain stable for long periods, so the ingredients must not react with one another chemically. Yet the paint must dry on the coated surface. Under some conditions some paints do not dry, which suggests that the drying process does not merely consist of evaporation of the liquid. In fact, there are three broad mechanisms of drying, two of which involve chemical reaction, while one does not.

### Drying without chemical reaction

In this case the paint *does* dry solely by evaporation of liquids. The polymer is fully formed in the can and, when free of solvent, is relatively hard and not sticky. During the drying process there is no chemical change in the polymer. If it was dissolved in the liquid solvents of the paint, then it remains soluble in those solvents. If it was carried as an emulsion or colloidal dispersion, it is not soluble in the liquid carrier, but it is soluble in other solvents, because it is a linear polymer. More is said about the sintering of particles of linear polymer into a continuous film in Chapter 11.

Nitrocellulose lacquers and decorative emulsion paints dry by this process. It is sometimes called 'lacquer drying'.

### Drying by chemical reaction

There are undoubtedly advantages in producing a dry paint film containing a cross-linked polymer as film former. Such polymers are, however, insoluble and so cannot be dissolved in solvents to form the vehicle of the paint in the can. We are therefore forced to have linear (or lightly branched) polymers, or even simple chemicals, in the can and to carry out a cross-linking chemical reaction *after the paint has been applied*. This can be done in two ways:

(i) DRYING BY CHEMICAL REACTION BETWEEN PAINT AND AIR. Oxygen and water vapour in particular are reactive chemical ingredients of the air. In Chapters 12 and 16, it will be shown how oxygen reacts with drying oils and other unsaturated compounds to produce free-radicals and bring about polymerization. In Chapter 15, we see how the reaction between water and isocyanates can cause condensation polymerization. Both reactions can give a cross-linked film. In both cases the principle is the same: the air is used as a chemical reactant and is kept apart from the reactive ingredients in the paint by the tightly fitting lid on the can. If it is not, it may cause the formation of a skin, which seals off the paint below from further reaction with the air. Alternatively, the skin may be permeable to the air and the reaction may spread right through the paint, turning it into a cross-linked polymer swollen with solvent. This is called an *irreversible gel* and of course at this point the paint becomes useless. However, assuming that the air is completely excluded, no reaction begins until the paint film is applied, presenting a large surface exposed to the air. As the solvents evaporate, cross-linking begins and the soft, sticky, low molecular weight, linear or branched polymers in the paint are converted to a hard, tough, cross-linked film, which will no longer dissolve in the solvents used in the original paint, or in any others.

This process is relatively slow at room temperature, chiefly because the reactive ingredients in the air must penetrate into the film before full hardening can occur. The paint hardens more rapidly if the reactive molecules in the paint are large. The effect of cross-linking is to build bigger and bigger molecules. Eventually the whole film could become one molecule. If the building 'bricks' (the original molecules) are large, less 'cement' (cross-links) will be needed to join them all together and the whole job (the hardening) will be done faster. However, as we have seen, there are advantages in keeping the molecular size down from the application and economy point-of-view. As a compromise, low molecular weight polymers (molecular weight 1000–5000) or large simple molecules are generally used. This leads to moderately high solids in the paint, decidedly a point in favour of cross-linking systems.

Chemical reaction with the air continues long after the paint film is apparently dry to touch and, in fact, the paint film will change its chemical nature slowly all the time it is in use. This means that the properties of the paint film change gradually also.

The principle of keeping the paint stable by having one reactant outside the can is quite general. This reactant need not be in the air: it could be in anything the paint comes into contact with during or after application, e.g. the application equipment or the surface being painted.

Exterior house paints derived from drying oils dry by reaction with the air.

(ii) DRYING BY CHEMICAL REACTION BETWEEN INGREDIENTS IN THE PAINT. Obviously the paint must remain chemically stable. The reactants must not react until the paint has been applied, yet (in this method) they must all be in the paint. This paradox is resolved, either by separating the reactive ingredients in two or more containers and mixing just before use, or by choosing ingredients which only react at higher temperatures or when exposed to radiation. The former method produces what is known as a 'two-pack paint'. Two-pack paints are less popular than their ready-mixed equivalents, because measuring is required before mixing and because of the limited period after mixing during which the paint remains usable (the 'pot life'). Sometimes two packs are avoided by so diluting the reactants with solvent, that reaction proceeds only slowly in the can, but much more quickly on a surface once the solvent has gone. Here the paint is not really stable, but a tolerable 'shelf life' is obtained. Whichever method is used, the chemical reactants may be sticky, low molecular weight polymers, or they may be simple chemicals. The reaction produces a cross-linked polymer.

Industrial stoving enamels and polyester wood finishes dry by chemical reaction of their ingredients.

A summary of the properties associated with the drying mechanisms is given in Table 1.

## Relative merits

It has been shown that paints are simple in outline, but complex in operation and formulation. How can the best choice of the many ingredients available be made? The choice of pigments and solvents is discussed in the next two chapters. This summary is concerned with the film-former.

Cross-linked or not cross-linked? There are points in favour of both alternatives. It is true that the hardest, toughest, most durable and solvent-resistant films contain cross-linked polymers. If they are to be matched in all respects except solvent resistance, linear polymers of high molecular weight must be used. This inevitably means much lower solids at application viscosity. Higher solids, without loss of molecular weight, can be obtained with emulsion systems, but these have their own special problems (see Chapter 11).

Against this can be argued the ease of polishing, pulling over or reflowing and the good patching properties of linear polymers. Also, a much simpler, fast method of drying, that is not particularly temperature sensitive and produces little or no handling or storage problem.

There is no right or wrong here. It simply depends on the job the paint has to do, the drying facilities available, the requirements of the user and whether the difficulties mentioned are slight or severe with the particular resins suitable for the job.

**Table 1**

| Method | Mol. wt. of film former in can | Solids | Type of polymer film | Ease of polishing patching and reflow | Rate of dry (no heat) | Minimum drying temperature | Handling and storage | Examples |
|---|---|---|---|---|---|---|---|---|
| (1) Evaporation | high | (i) low, 10–35% (solution) (ii) medium-high, 40–70% (emulsion) | linear | good | fast | no practical limit (solutions) | good | Nitrocellulose and other lacquers; some emulsion paints; some organosols |
| (2) Chemical reaction between paint and air | low | medium to high, 35–100% | cross-linked | fair–poor | slow–moderate | very slow in cold weather | cans must be well sealed | Decorative paints; some stoving enamels; one-pack polyurethanes |
| (3) Chemical reaction between paint ingredients | low or very low | medium to high 30–100% | cross-linked | fair–poor | fairly fast | varies; 10–15°C common | two-pack or short shelf life, unless stoving or radiation curing type | Industrial stoving finishes; acid-catalysed, poly-urethane and polyester wood finishes |

# Eight

# Pigmentation

The outline formula of a paint has been given in Chapter 7 and the nature of polymers and resins has been described in Chapter 5. In the next three chapters, the other main ingredients of surface coatings will be covered briefly, before particular paints are described in the remainder of this book.

A detailed study of the chemistry of pigments is beyond the scope of an introductory book such as this and is already available elsewhere (see Appendix). Instead, this Chapter reviews some of the key properties of pigments and discusses their selection and methods of incorporating them into paint.

### Pigment properties

Any surface coating pigment may be asked to carry out some (or perhaps, all) of the following tasks:

(1)  To provide colour.
(2)  To obliterate previous colours.
(3)  To improve the strength of the paint film.
(4)  To improve the adhesion of the paint film.
(5)  To improve the durability and weathering properties.
(6)  To increase the protection against corrosion.
(7)  To reduce gloss.
(8)  To modify flow and application properties.

To choose a pigment to carry out a given selection of these eight functions, we must know about the following properties of the pigment

|  |  |
|---|---|
| (i)  Tinting strength | (vi)  Particle size |
| (ii)  Lightfastness | (vii)  Particle shape |
| (iii)  Bleeding characteristics | (viii)  Specific gravity |
| (iv)  Hiding power | (ix)  Chemical reactivity |
| (v)  Refractive index | (x)  Thermal stability |

(i) TINTING STRENGTH. The majority of paints contain white pigment, which is tinted to the appropriate pastel or mid-shade with coloured pigments. If a lot of coloured pigment is required to achieve the shade, it is said to have poor tinting strength. The tinting strength of a coloured pigment is related to that of some standard pigment of similar hue. If numerical values are given, then

$$
\begin{array}{l}
\text{amount of alternative} \\
\text{pigment required to} \\
\text{produce the shade}
\end{array}
= 
\begin{array}{l}
\text{amount of standard} \\
\text{pigment required}
\end{array}
\times 
\frac{\text{Tinting strength of standard}}{\begin{array}{l}\text{Tinting strength of}\\\text{alternative pigment}\end{array}}
$$

The tinting strength of a pigment is independent of its hiding power, since the comparison of shades is done at film thicknesses that completely hide the substrate. Relatively transparent pigments can have high tinting strengths.

The phrase 'tinting strength' is sometimes applied to white pigments. A single coloured pigment is used for the comparison of white pigments at a fixed shade.

(ii) LIGHTFASTNESS. To give a good colour intially is not enough. The colour must last, preferably as long as the paint film. Many pigments fade or darken or change shade badly in the light. This is because the ultra-violet rays in sunlight are sufficiently energetic to break certain chemical bonds and thus change molecules. A change in chemical structure means a change in the ability to absorb light in the visible region of the spectrum, with consequent loss of colour or variation of hue. On the other hand, if the pigment can absorb ultra-violet rays without breakdown, it will protect the binder. The energy is dissipated harmlessly as heat.

(iii) BLEEDING CHARACTERISTICS. Not all pigments are completely insoluble in all solvents. Colour in the right place is all very well, but when white lettering applied over a red background turns pink, it is not so good. What happens is that the solvents of the white paint dissolve some of the red pigment in the background coat and carry it into the white overlayer. Organic reds as a group are particularly prone to this fault (though theoretically it can occur with any colour), so it has been given the name 'bleeding'.

(iv) HIDING POWER. Ideally one coat of paint should obliterate any colour. Frequently two are needed, but in any case the total thickness of paint applied should be no more than that necessary for the required degree of protection of the surface and the presentation of a smooth, pleasing appearance. Such thicknesses are suprisingly small, usually $25-100\,\mu$m. To obliterate, the pigments used must prevent light from passing through the film to the previous coloured layer and back to the eye of an observer. The pigments do this by absorbing and scattering the light. The *hiding power* of a paint is expressed as the number of square metres covered by one litre of paint to produce complete hiding and the hiding power of a pigment as the number of square metres covered per kg of pigment, which has been dispersed in a paint and applied so that it will just hide any previous colour. Hiding power depends upon the wavelengths and total amount of light that a pigment will absorb, on its refractive index and also on particle size and shape.

(v) REFRACTIVE INDEX. Chapter 6 showed how light rays can be bent and how they may be returned to the eye from a paint film. The light rays suffer refraction, diffraction and reflection by transparent particles that have a refractive index differing from that of the film in which they lie. White pigments are transparent in large lumps, but white in powder form, because they have high refractive indices (2·0–2·7). These are greater than the refractive indices of film-formers (1·4–1·6), so the pigments give a film its white colour by the mechanism described for incompatible resins, in Chapter 6. Titanium dioxide ($TiO_2$) pigments, in particular, have excellent hiding power, because their refractive indices are so much higher than those of film-formers and because they have the optimum particle size. Extender pigments are also transparent in bulk and white as powders, but they do not colour paints, because their refractive indices scarcely differ from those of film-formers.

(vi) PARTICLE SIZE. There is an ideal particle diameter for maximum scattering of light at interfaces and this is approximately equal to the wavelength of the light *in the particle*. As a rough guide, the optimum diameter is approximately half the wavelength of the light in air, i.e. about 0·2–0·4 $\mu$m. Below this size, the particle loses scattering power; above it, the number of interfaces in a given weight of pigment decreases. The hiding power of a transparent pigment is reduced. Pigments have particle diameters varying from 0·01 $\mu$m (carbon blacks) to approximately 50 $\mu$m (some extenders). No sample of pigment contains particles all of an identical size; rather there is a mixture of sizes with an average diameter.

Connected with particle size are *surface area* and *oil absorption*. If a cube of pigment is cut in two, the weight of pigment is the same, the number of particles is doubled, the size of particle is halved and two new surfaces are formed along the cut, in addition to all the original surface of the cube. Thus in any fixed weight of pigment, the smaller the particles, the larger the area of pigment surface.

It is interesting to note that, while 1 gram of rutile titanium dioxide white (particle diameter 0·2–0·3 $\mu$m) has a surface area of 12 $m^2$, 1 gram of fine silica (particle diameter 0·015–0·2 $\mu$m) has a surface area of 190 $m^2$ – about the area of a singles court at tennis!

The surface area is indicated by the oil absorption, which is the minimum weight in grams of a specified raw linseed oil that is required to turn 100 g of pigment into a paste. The oil is added slowly to the pigment with thorough mixing and shearing between the walls of the vessel and a rod. While the oil is in the process of replacing air molecules on the particle surfaces (this is called *wetting* the pigment), the pigment/oil mixture remains a crumbly mass. When wetting is as complete as it can be, the next additions of oil will fill the spaces between the particles. As soon as there is a slight oil surplus, the mixture becomes a paste, because the particles are free to flow past one another, as a result of the lubricating action of the oil. It can be seen that, while the oil absorption value is related to the surface area, it depends on how successful the operator is in displacing all the air from the whole surface. With many pigments he will not be at all successful, using such rudimentary equipment. Some surfaces will have more attraction for linseed oil molecules than others. Obviously the greater the area of pigment left unwetted, the lower the oil absorption will appear to be. If the volume of space between the particles varies, as it will do from pigment to pigment according to the particle shape and distribution of sizes, then this will also affect the final figure. At best oil absorption is a rough guide to surface area, but its merit is that it is easily and quickly measured and gives the paint formulator some idea of the type of pigment that he is dealing with.

The most important part of a pigment particle is its surface. At the surface are those chemical groups that will make contact with the chemical groups in the varnish. These surface groups will determine whether the pigment has any attraction for varnish molecules or not (whether the pigment is easily wetted) and whether the pigment has *special* attractions for *parts* of the varnish (e.g. drier molecules, see Chapter 12). This latter feature may be undesirable, since molecules firmly attached to pigment surfaces are not free to move about and carry out their functions in the varnish. If they are molecules of a paint additive, they are bound to be present in small quantities; a pigment of very large surface area could totally absorb the additive, rendering it ineffective. If these groups on the pigment surface attract *each other* strongly, the pigment particles will tend to cluster, resisting wetting and dispersion and setting up a loosely bound 'structure' of particles in the paint, which will affect its application properties (see Chapter 10). This phenomenon is called 'flocculation'. If the chemical nature of the pigment surface is not known, it is a general rule that the larger the surface area, the more active the pigment surface will appear to be.

(vii) PARTICLE SHAPE. Particles may be nearly spherical, cubic, nodular (a rounded irregular shape), acicular (needle- or rod-like) or lamellar (plate-like). Since particle shape affects pigment packing, it therefore affects hiding power. Rod-shaped particles can reinforce paint films, like iron bars in concrete, or they may tend to poke through the surface reducing gloss. Such rough surfaces may help the next coat to stick more easily, so this type of pigment could be useful in an undercoat. Plate-shaped particles tend to overlap one another like tiles on a roof, making it more difficult for water to penetrate the film. Aluminium and micas pigment have this shape.

(viii) SPECIFIC GRAVITY. This is the weight of a substance in grams divided by its net volume in millilitres (excluding, for example, volume occupied by air between pigment particles). For rutile titanium dioxide white, this value is 4·1; for white lead, 6·6. Since pigment is sold to the paint maker by the kilo and is then resold to his customers by the litre, the specific gravity of the pigment is important. An expensive pigment (per kg) may yet prove economical if its specific gravity is low – a little goes a long way. Extender pigments are not only cheap, but have low specific gravities: that is why they are used to increase pigment volume, where the hiding power of the coloured pigment is good enough at low concentrations.

(ix) CHEMICAL REACTIVITY. Chemical reactivity can make some pigments unsuitable for some purposes. For example, zinc oxide is amphoteric and should not be used as a white pigment with resins containing a high proportion of acid groups. Soap formation will occur and – because zinc is divalent – this will tend to cross link the resin, causing excessive viscosity increase on storage. The paint is said to become 'livery' and unusable.

On the other hand, some pigments are included particularly *because they are reactive*. Excellent examples are the *anti-corrosive pigments*. Especially effective are the chromates of zinc, lead and strontium. When the paint film is permeated by water, these pigments slowly release chromate ions, because of their low but measurable solubilities. The anti-corrosive action of chromates is discussed in some detail in Chapter 17. Another long-established anti-corrosive pigment is red lead, which works well in oil- or alkyd-based paints (Chapter 12). It is though that this is due to the formation of lead soaps of azelaic acid, $HOOC.(CH_2)_7. COOH$, one of the aging by-products of unsaturated oils. Both lead and chromate pigments are relatively toxic and the search is on for less toxic alternatives that are as effective. Nothing with the all-round effectiveness of chromates has yet been found, but some success has been had with phosphate pigments and with silica pigments which release calcium on an ion-exchange basis.

These examples illustrate the advantages of knowing the pigment chemistry. This is especially necessary with the newer (probably organic) pigments, whose precise chemical nature may not be disclosed by the manufacturer. Even with the more traditional pigments, other ingredients may be added by the pigment manufacturer to alter the crystal shape, or to provide a coating on the pigment surface that will make the pigment easier to disperse. Unless precise knowledge can be obtained, the paint formulator is very much in the hands of the pigment manufacturer and must rely upon his literature.

(x) THERMAL STABILITY. The temperature at which a pigment decomposes or alters its nature (e.g. melts) can be very important if the pigment is required for a paint to be stoved at high temperatures, or if the paint is to be heat-resistant.

Where is all this information to be obtained? Partly in general pigment and paint literature (where traditional pigments and well-defined groups of pigments are concerned), partly from the technical literature supplied by the pigment manufacturer and partly by experiment. For a reference to experimental methods for determining pigment properties, see the Appendix.

## Pigment types

Whole books have been written on pigments classified by chemical types. Here we have room only to attempt a much broader classification and to say a few words about the implications of the classification. All pigments are either:

(a) Natural or synthetic, and
(b) Organic or inorganic chemicals

### (a) Natural and synthetic pigments

It is unlikely that there is any organic pigment, the naturally occurring form of which is still in industrial use today. Many inorganic pigments, however, are still dug out of the earth's crust, crushed, washed and graded by size. Frequently, there is a synthetic equivalent – that is to say, a pigment made from other ingredients by a chemical process – apparently the same chemically, but often different in properties. The differences arise because:

(1) whereas the natural pigment is available in the crystal form found in nature, the synthetic product may be induced to take a more desirable crystal shape.
(2) the natural product may be contaminated by some impurity, such as silica, which it is uneconomic to remove: the synthetic product should be pure or nearly so.
(3) crushing produces a wide range of particle sizes and grading methods may not be able to remove all oversize or undersize particles from the desired range. A pigment produced by precipitation under controlled conditions, should have a much more uniform size range.

The chief family of pigments in which natural varieties are still of importance, is the iron oxide family: ochres, umbers and siennas; red, yellow and black iron oxides.

### (b) Organic and inorganic pigments

There are now far more organic pigments than inorganic ones, though some of the newest contain both metallic (inorganic) elements and organic structures. Again, many organic pigments are organic chemicals deposited on an inorganic (e.g. aluminium hydroxide) 'core'. These pigments are called 'lakes'. It is difficult to lay down hard and fast rules, as there are exceptions in every case, but for some properties there is a clear-cut advantage for the bulk of one type of pigment over the majority of the other type. These properties are listed below:

| Property | Preference | Reasons |
| --- | --- | --- |
| (1) Brilliance and clarity of hue | Organic | The most attractive, cleanest colours can only be obtained with organic pigments. |
| (2) White and black paints | Inorganic | The purest white pigment is titanium dioxide and the most jet black, carbon (usually considered inorganic). There are no organic blacks and whites. |
| (3) Non-bleeding | Inorganic | Inorganic compounds have negligible solubilities in organic solvents. Some organics are very insoluble. |
| (4) Lightfastness | Inorganic | The valency bonds in inorganic compounds are generally more stable to ultra-violet light than those in organic compounds. |
| (5) Heat stability | Inorganic | Very few organic compounds are stable at or above 300 °C. Some decompose or melt at much lower temperatures. |
| (6) Anti-corrosive action | Inorganic | All anti-corrosive pigments are inorganic. |

This list may give the impression that it is best to avoid organic pigments, but factor (1) is a very powerful argument for using them. The very best organic or organometallic pigments give the highest standard of performance for most uses, e.g. the phthalocyanine ('Monastral') pigments, the quinacridone ('Cinquasia') pigments, the perylene and perinone pigments, the dioxazine pigments and the high molecular weight azo pigments.

## Pigment selection

To sum up, the paint formulator's method of pigment selection is as follows:

(1) Examine a pattern of the colour to be produced in paint. Estimate the number of different hues that will have to be blended to produce the colour. A suitable pigment has to be found to provide each hue.
(2) Define the properties required from the pigments.
(3) Select a suitable pigment or pigments in each hue. In order to obtain pigments with the required properties, consult the following sources where necessary:
(a) General paint and pigment literature. There are traditional pigments for particular uses and binders.
(b) *The Colour Index* (Society of Dyers and Colourists, Bradford, Yorks.). The properties of pigments are given, together with the names of commercial grades and their manufacturers.
(c) Pigment manufacturers' literature. Examples of colour may be given.
(4) Match the colour (see below) with one or more pigment combinations.
(5) Test the paints produced.

## Dispersion

The next stage is to get the pigment into the paint. The pigment is usually supplied as a powder, in which the granules are actually aggregates of the fine particles produced by the pigment manufacturer. These particles must be dispersed or separated from one another and evenly distributed throughout the paint as a colloidal suspension. For this suspension to have the maximum stability in organic solvents, the surface of each particle should be completely wetted with the varnish: there should be no intervening layers of air or adsorbed water.

Wetting with solvent alone is not enough and pigment dispersions in solvent have poor stability. Each particle in the pigment suspension must be stabilized by polymer chains, anchored to its surface by intermolecular attractions, yet extending out into the varnish because of their attractions for the solvent molecules. When two such particles approach one another, they do not adhere. Contact between them involves the intermingling of polymer chains from the two particles. Locally, in the region between the particles, the concentration of resin molecules is higher than at other points in the varnish. This upsets the equilibrium of the system, so solvent molecules diffuse into this region, dilute the concentration and restore equilibrium. The osmotic pressure involved is sufficient to separate the particles.

A resin suitable for dispersion usually contains polar groups (which provide the attraction for pigment surface molecules) and is completely soluble in the solvent mixture of the dispersion. Sometimes, surfactants (Chapter 10) are used to bridge the particle-resin interface and assist wetting. In water, ionic surfactants (e.g. soaps) can provide the pigment surface with an electrical charge. The particles, being of like charge, repel one another and the dispersion is stable.

Dispersion is usually carried out in a *mill*, a machine in which the aggregates are subjected to the forces of shear and (sometimes) attrition. When *shear* is the dispersing action, the aggregates are squeezed between two surfaces moving in opposite directions, or in the same direction but at different speeds. It is just like making cocoa, where the powder has to be dispersed in a little milk to form a paste. Dispersion is carried out by a

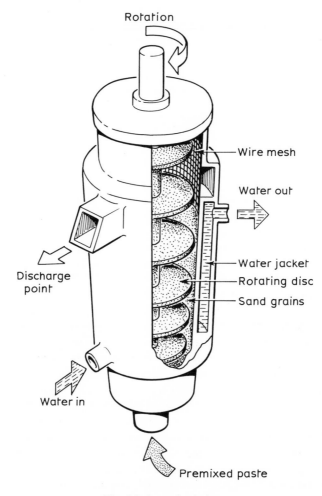

**Fig. 24** A sand grinder.

shearing motion between spoon and cup. In paint mills, the principle is the same, but the power and degree of dispersion are much greater.

During the manufacture of the pigment, the particles are reduced to their ultimate size by crushing in a liquid slurry. Where *attrition* is part of a dispersion process, the conditions are much milder and there is no attempt to fracture individual particles. In the viscous medium of the varnish, aggregates – not particles – are broken.

Several types of mills are used. The *high speed disperser* is used for easily dispersed pigments and consists of a horizontal disc with a serrated edge, which rotates at high speed about a vertical axis. The *ball mill* is a cylinder revolving about its axis, the axis being horizontal and the cylinder partly filled with steel or steatite (porcelain) balls, or pebbles. The speed of rotation is such that the balls continually rise with the motion and then cascade down again, crushing and shearing the pigment. In the *sand grinder* (Fig. 24) or the *bead mill*, the axis of rotation is vertical, the grinding medium is coarse sand or glass beads and the charge is induced to rotate at higher speeds by revolving discs in a stationary container. The ball mill produces a batch of pigment dispersion; the sand grinder gives a continuous output of dispersed pigment.

*Roller mills*, which are rapidly disappearing from the paint industry, consist of a number of horizontal steel rolls placed side-by-side and moving in opposite directions, often at different speeds, with very small clearances

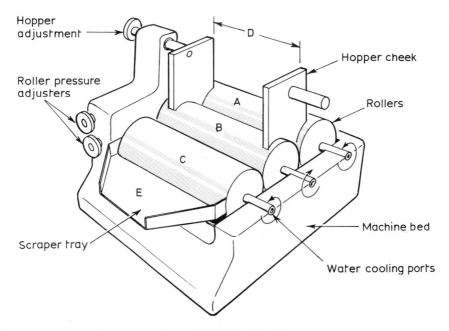

**Fig. 25** A triple roll mill.

in between. In these gaps, the pigment/varnish mixture is sheared. The *triple roll* is shown in Fig. 25. A paste is fed in at D, is lightly sheared between rollers A and B and more severely between B and C. A scraper blade E removes the dispersion. All rollers are independently driven and run at different speeds to increase shear. A fourth type of mill is the *heavy duty* or *'pug' mixer*, in which roughly S-shaped blades (Fig. 26) revolve in opposite directions and at different speeds in adjacent troughs (Fig. 27). A stiff paste is required. In modern versions of this type of mixer, the blades are mounted vertically and dip into the dispersion vessel (a vertical cylinder) like the blades of a kitchen dough mixer.

**Fig. 26** Blade for a heavy duty mixer.

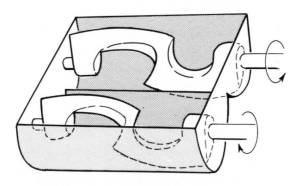

**Fig. 27** Troughs of a heavy duty mixer.

Several alternative mills are available, which the reader may discover elsewhere (see Appendix). These few examples are quoted to illustrate the principles involved.

It is obvious from the mention of stiff pastes that the whole paint is not charged into the mill. In fact, the paint maker aims to put in the maximum amount of pigment and the minimum amount of varnish to get the largest possible paint yield from his mill. This mixture forms the grinding or *first stage*. When the dispersion is complete (after a period varying from 10 minutes to 48 hours according to the materials and machinery involved), the consistency is reduced with further resin solution or solvent, so that the mill can be emptied as cleanly as possible. This is the 'let-down' or *second stage* and may take up to two hours. The *third* or final *stage* (carried out in a mixing tank) consists of the completion of the formula by addition of the remaining ingredients. A break-down of a possible ball mill formula looks like this:

|  | wt % |  |
|---|---|---|
| Pigment | 10·0 | ⎫ |
| Resin | 1·0 | ⎬ Stage I (grinding). Then add: |
| Solvent | 3·0 | ⎭ . . . . . . . . . . . . . . . |
| Resin | 1·0 | ⎫ Stage II (let down). Empty mill – then add: |
| Solvent | 3·0 | ⎭ . . . . . . . . . . . . . . . |
| Resin | 29.0 | ⎫ |
| Solvent | 51·5 | ⎬ Stage III (completion of formula) |
| Additives | 1·5 | ⎭ |
|  | 100·0 |  |

The exact composition of Stage I is found by experiment, to give the minimum grinding time and the most stable and complete dispersion. Stages II and III also require care, as hasty additions in an incorrect order can cause the pigment to re-aggregate (flocculate).

The amount of pigment in the formula is that required for the appropriate colour, hiding power, gloss, consistency and durability. As a rough guide, the amount might vary from one third of the binder weight to an equal weight (for a glossy pastel shade).

## Colour matching

The colour of a paint is usually matched to some colour pattern. It must, at any rate, be matched to the colour of the previous batch of the same paint. If a new shade is to be matched in a paint, the procedure for matching is as follows:

(1) The main hue and the undertones are observed; also whether the colour is 'clean' or 'dirty.

(2) From experience and by application of the principles of subtractive mixing, the pigments likely to give the colour are chosen. The procedure is described under **Pigment Selection.**

(3) Each one of these pigments is dispersed separately and each dispersion is converted to a paint. The colour matcher now has several paints, each containing only one pigment. These are known as *colour solutions*.

(4) The colour solutions are blended, the choice of solutions and proportions being varied, until finally the correct blend to give the colour is obtained. The proportions of pigment in this blend are recorded.

(5) The final paint is often made by blending colour solutions, but frequently a more stable dispersion is obtained if all the pigments are
dispersed simultaneously in the same mill. This is called *co-grinding* the
pigments. If the colour is then slightly off shade, it is adjusted by
additions of colour solutions or concentrated pigment dispersions.

In larger paint companies, where spectrophotometers or colorimeters and
a computer are available, these stages can often be speeded up considerably.
The new shade can be measured by the colorimeter and expressed precisely
as a group of numerical values. This information can be passed to the
computer, which has been previously programmed with the values for
numerous pigments and with the 'rules' governing the mixing of these
pigments to give the desired colour. The computer will then print out a
choice of pigmentations (type and proportions) for achieving the colour.
The formulator makes a selection and will usually find that the colour he
produces to the computer's instructions will be close to the desired colour
and that the recipe will require only minor modifications to give the exact
shade.

Most colours, even dark ones, require substantial proportions of white
pigment. Clean colours require organic pigments and should be matched
with the minimum number of pigments. The subtractive effect increases
with the number of pigments, reducing the reflected light intensity and
hence 'dirtying' the colour. If darkening or 'dirtying' of a colour is required,
it can be done with black pigment. The use of black pigment can markedly
alter the shade. Thus brown is a mixture of yellow (or red) and black.

# Nine

# Solvents

Before we start, it should be explained that the word 'solvent' is used loosely in the paint industry to denote any liquid that may be used in a paint: it tells you nothing about the liquid's ability to dissolve the polymers in the paint. In this chapter, we will have to use the word in the paint sense and in the scientific sense (as defined in Chapter 1). It is hoped that the meaning intended will be clear from the context and, to minimize the confusion, the word 'liquid' will be used as much as possible, where no description of solvent power is intended.

## Chemical types

The following liquids are commonly used in paints to carry the pigment and binder:

  (i) Water
 (ii) Aliphatic hydrocarbon mixtures: chiefly paraffins
(iii) Terpenes
 (iv) Aromatic hydrocarbons
  (v) Alcohols
 (vi) Esters
(vii) Ketones
(viii) Ethers and ether-alcohols
 (ix) Nitroparaffins
  (x) Chloroparaffins

(i) WATER is the main ingredient of the continuous phase of most emulsion paints (Chapter 11). It is also used alone, or blended with alcohols or ether-alcohols, to dissolve water-soluble resins (e.g in water-reducible paints) or dyestuffs (e.g. in water stains).

Any type of resin can be made water-soluble. The usual procedure is to incorporate sufficient carboxyl groups into the polymer to give the resin a high acid value (Chapter 12) of 50 or more. These groups are then neutralized with a volatile base, such as ammonia or an amine, whereupon the resin becomes a polymeric salt, soluble in water or water/ether-alcohol mixtures. For example, drying oils can be made acidic if they are heat-bodied in the presence of maleic anhydride, reaction taking place between the double bonds in both materials (see **Bodied Oils,** Chapter 12). If the carboxyl groups are not fully neutralized, an emulsion can be produced.

The polymeric part of the salt is negatively charged and can be attracted to and discharged at the anode (see Chapter 2) in an electrolytic cell. If the anode is a metallic object, say a car body, then the discharged resin (or paint, if it is associated with pigment particles) will deposit on that object and coat it. This is called *anodic electrodeposition.* If the paint film deposited is not porous then, as the film thickness increases, so does its electrical resistance (or insulating power) and thus the deposition rate slackens and deposition eventually stops. This has the effect of automatically controlling film thickness and of encouraging deposition at points which are more thinly coated. From this arise two of the main advantages of electrodeposition: it can be a fully automated process and suitably formulated paints (with good 'throwing power') can be made to coat all parts of an article of any shape, however awkward, with a uniform paint coating.

Instead of acidic carboxyl groups, resins can be made to include basic, e.g. amine, groups, and these can be neutralized with relatively volatile acids, e.g. acetic or lactic acids, to produce resin salts soluble in water or water-solvent mixtures. Again, incomplete neutralization produces an emulsion. In these systems, the resin is positively charged and can be discharged at a cathode (see Chapter 2). This is the basis of *cathodic electrodeposition.*

Even if the polymeric salts are capable of anodic or cathodic electrodeposition, waterborne paints made from them do not have to be applied in this way. The solution or emulsion paints produced can be applied by virtually any technique. There will be an increasing use of this type of paint as legislation, economics and world petroleum shortages encourage the use of decreasing amounts of volatile organic solvents and paints of low flammability.

An alternative method of obtaining water-solubility, without salt or ion formation, is to incorporate short lengths of naturally water-soluble polymer into the main polymer structure. Often polyethylene glycol is used. This is a polyether (see Chapter 15) of molecular weight around 1000, which is made from ethylene oxide and ethylene glycol. 5–20 per cent by weight is incorporated into the resin, usually by esterification. The resin thus modified can be emulsified in water by simple stirring, or can be dissolved in water/ether-alcohol mixtures. Whereas ammonia will evaporate from the paint film eventually, polyethylene glycol will not, so the film remains permanently sensitive to water.

The chief virtues of water are its availability, cheapness, lack of smell, non-toxicity and non-flammability. However, it is not an ideal paint liquid, because of its limited miscibility with other liquids and because the film-formers designed to be dissolved or dispersed in it, usually remain permanently sensitive to it. In fact, its abundance in nature makes it any paint film's worst enemy.

(ii) ALIPHATIC HYDROCARBONS are usually supplied as mixtures, because of the difficulty of separating the individual compounds. A boiling range for the mixture is usually quoted, e.g. S.B.P. (special blend of petroleum) spirit No. 3, 98–122 °C. Many mixtures also contain a percentage of aromatic hydrocarbons, e.g. white spirit, 155–195 °C, about 15 per cent aromatic.

(iii) TERPENES commonly used are turpentine, dipentene and pine oil. Turpentine varies with grade, but is principally $\alpha$-pinene; dipentene is mainly limonene; while pine oils are mixtures, chiefly of terpene alcohols. Turpentine was once the main solvent for house paints, but has now been replaced by white spirit. The terpenes can be used as anti-skinning agents (Chapter 10).

(iv) to (vii) the AROMATIC HYDROCARBONS, ALCOHOLS, ESTERS and KETONES supplied to the paint industry are fairly pure named compounds. Some aromatic mixtures are sold cheaply under proprietary names.

(viii) ETHERS are not often used, but ether-alcohols – which contain both the ether (—C—O—C—) and alcohol (—C—O—H) groups – are very common, e.g.

$CH_3(CH_2)_3 \cdot O \cdot CH_2 \cdot CH_2OH$
ethylene glycol monobutyl
ether(2-butoxy ethanol)

$CH_3 \cdot O \cdot CH_2 \cdot CHOH \cdot CH_3$
propylene glycol monomethyl
ether (1-methoxy propan-2-ol)

$HO \cdot CH_2 \cdot CH_2 \cdot O \cdot CH_2 \cdot CH_2 \cdot OH$
diethylene glycol (DEG)

$CH_3 \cdot O \cdot CH_2 \cdot CH_2 \cdot O \cdot CH_2 \cdot CH_2 \cdot OH$
DEG monomethyl ether

(ix) and (x) NITRO- AND CHLOROPARAFFINS are uncommon solvents in paints, the latter because they are somewhat toxic, the former because of cost and some recent evidence of toxicity.

## Solvent properties

The most important properties of a liquid for paints are:

(a) Solvency, i.e. whether it is a solvent or non-solvent for a given film-former. This depends on the film-former and is not an independent property of the liquid.
(b) Viscosity or consistency.
(c) Boiling point and evaporation rate.

**Table 2** Solvent properties

| Solvent | Formula | Solvency | | Viscosity at 20°C (centipoises) | Boiling point (°C) | Flash point (closed cup) (°C) |
| --- | --- | --- | --- | --- | --- | --- |
| | | H-bonding group | Solubility parameter | | | |
| Water | $H_2O$ | III | 23·4 | 1·002 | 100 | None |
| **Aliphatic hydrocarbons** | | | | | | |
| cyclo-hexane | 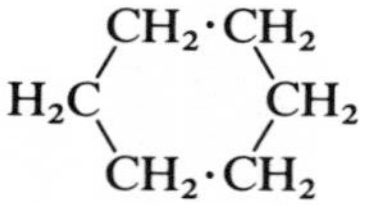 $CH_2 \cdot CH_2$ $H_2C$ $CH_2$ $CH_2 \cdot CH_2$ | I | 8·2 | 0·89* | 81 | 3 |
| White spirit | | I | | | 155–195 | 33[m] |
| Odourless white spirit | | I | 6·9 | | 180–207 | 55 |
| **Terpenes** | | | | | | |
| Dipentene | 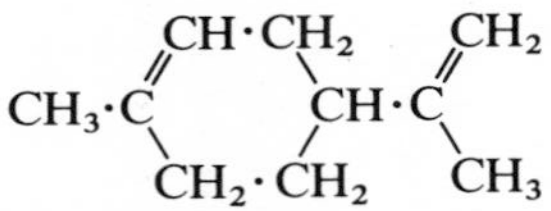 $CH \cdot CH_2$ $CH_2$ $CH_3 \cdot C$ $CH \cdot C$ $CH_3$ $CH_2 \cdot CH_2$ | I | 8·5 | 0·975* | 175–190 | 32[m] |
| Turpentine | 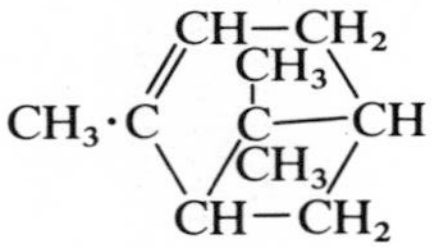 $CH-CH_2$ $CH$ $CH_3 \cdot C$ $C$ $CH_3$ $CH-CH_2$ | I | 8·1 | 1·26* | 150–170 | 33 |
| Pine oil | | I | 8·6 | 6–26* | 195–220 | 74–88[op] |

**Aromatic**
**hydrocarbons**

| | | | | | | |
|---|---|---|---|---|---|---|
| Toluene | $C_6H_5 \cdot CH_3$ | 8·9 | I | 0·55* | 111 | 4 |
| Xylene | $C_6H_4 \cdot (CH_3)_2$ | 8·8 | I | 0·586 | 138–144 | 27 |
| Styrene | $C_6H_5 \cdot CH{=}CH_2$ | 9·3 | I | 0·77* | 146 | 31 |
| Vinyl toluene | $CH_3 \cdot C_6H_4 \cdot CH{=}CH_2$ | | I | | 164–170 | 32 |
| **Alcohols** | | | | | | |
| Methanol | $CH_3OH$ | 14·5 | III | 0·547 | 65 | 12–14 |
| Ethanol | $C_2H_5OH$ | 12·7 | III | 1·200 | 78 | 14 |
| n-Propanol | $CH_3 \cdot (CH_2)_2 \cdot OH$ | 11·9 | III | 2·25 | 97 | 15 |
| iso-Propanol | $(CH_3)_2 \cdot CH \cdot OH$ | 11·5 | III | 2·15* | 82 | 12 |
| n-Butanol | $CH_3 \cdot (CH_2)_3 \cdot OH$ | 11·4 | III | 2·948 | 118 | 35 |
| iso-Butanol | $(CH_3)_2 \cdot CH \cdot CH_2 \cdot OH$ | | III | 3·9* | 108 | 25 |
| sec-Butanol | $CH_3 \cdot CH_2 \cdot CH(CH_3) \cdot OH$ | 10·8 | III | 3·15 | 100 | 24 |
| cyclo-Hexanol | $\begin{array}{ccc} & CH_2 \cdot CH_2 & \\ H_2C & & CH \cdot OH \\ & CH_2 \cdot CH_2 & \end{array}$ | 11·4 | III | 81† | 162 | 68 |
| Ethylene glycol | $HO \cdot CH_2 \cdot CH_2 \cdot OH$ | 14·2 | III | 17* | 198 | 111 |
| Glycerol | $CH_2OH \cdot CHOH \cdot CH_2OH$ | 16·5 | III | 494§ | 290 | 160 |

* at 25°C  † at 38°C  § at 26.5°C  ‡ at 15°C  m = minimum  op = open cup

**Table 2** Solvent properties (*continued*)

| Solvent | Formula | Solvency | | Viscosity at 20°C (centipoises) | Boiling point (°C) | Flash point (closed cup) (°C) |
|---|---|---|---|---|---|---|
| | | H-bonding group | Solubility parameter | | | |
| **Esters** | | | | | | |
| Methyl acetate | $CH_3 \cdot CO \cdot O \cdot CH_3$ | II | 9·6 | 0·38* | 57 | −9 |
| Ethyl acetate | $CH_3 \cdot CO \cdot O \cdot C_2H_5$ | II | 9·1 | 0·455 | 77 | −4 |
| Butyl acetate | $CH_3 \cdot CO \cdot O \cdot C_4H_9$ | II | 8·5 | 0·671* | 127 | 23 |
| Methoxypropyl acetate | $CH_3 \cdot CO \cdot O \cdot CH(CH_3) \cdot CH_2 \cdot O \cdot CH_3$ | II | 9·2 | 1·2 | 140–150 | 46 |
| **Ketones** | | | | | | |
| Acetone | $CH_3 \cdot CO \cdot CH_3$ | II | 10·0 | 0·316* | 56 | −17 |
| Methyl ethyl ketone | $CH_3 \cdot CO \cdot C_2H_5$ | II | 9·3 | 0·423‡ | 80 | −4 |
| Methyl *iso*-butyl ketone | $CH_3 \cdot CO \cdot CH_2 \cdot CH(CH_3)_2$ | II | 8·4 | 0·546* | 116 | 16 |
| *cyclo*-Hexanone | 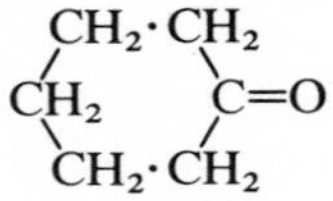 $CH_2 \cdot CH \genfrac{}{}{0pt}{}{CH_2 \cdot CH_2}{CH_2 \cdot CH_2} C{=}O$ | II | 9·9 | 1·94 | 157 | 47 |
| Methyl *cyclo*-hexanone | 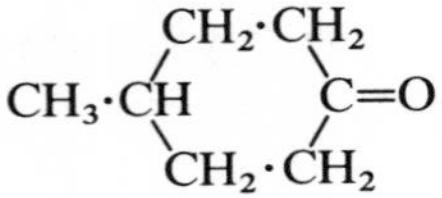 $CH_3 \cdot CH \genfrac{}{}{0pt}{}{CH_2 \cdot CH_2}{CH_2 \cdot CH_2} C{=}O$  and isomers | II | 9·3 | 1·75 | 165–175 | 47 |

**Ethers and ether-alcohols**

| | | | | | | |
|---|---|---|---|---|---|---|
| Diethyl ether | $C_2H_5 \cdot O \cdot C_2H_5$ | II | 7·4 | 0·233 | 35 | −40 |
| 1-Methoxy propan-2-ol | $CH_3 \cdot O \cdot CH_2 \cdot CHOH \cdot CH_3$ | II | 10·2 | 1·65* | 120 | 38[op] |
| 1-Ethoxy propan-2-ol | $CH_3 \cdot CH_2 \cdot O \cdot CH_2 \cdot CHOH \cdot CH_3$ | II | 9·0 | 1·68* | 132 | 43[op] |
| 2-Butoxy ethanol | $C_4H_9 \cdot O \cdot CH_2 \cdot CH_2 \cdot OH$ | II | 8·9 | 3·318* | 171 | 61 |
| Diethylene glycol (DEG) | $HO \cdot CH_2 \cdot CH_2 \cdot O \cdot CH_2 \cdot CH_2 \cdot OH$ | III | 9·1 | 30* | 245–250 | 124 |
| DEG monomethyl ether | $CH_3 \cdot O \cdot CH_2 \cdot CH_2 \cdot O \cdot CH_2 \cdot CH_2 \cdot OH$ | II | 9·6 | 3·53* | 194 | 93[op] |

**Nitroparaffins**

| | | | | | | |
|---|---|---|---|---|---|---|
| Nitromethane | $CH_3 \cdot NO_2$ | I | 12·7 | 0·62 | 101 | 35 |
| Nitroethane | $C_2H_5 \cdot NO_2$ | I | 11·1 | 0·62 | 114 | 28 |
| 1-Nitropropane | $CH_3 \cdot CH_2 \cdot CH_2 \cdot NO_2$ | I | 10·7 | 0·81 | 132 | 34[op] |

**Chlorinated paraffins**

| | | | | | | |
|---|---|---|---|---|---|---|
| Methylene chloride | $CH_2Cl_2$ | I | 9·7 | 0·425* | 41 | None |
| Ethylene dichloride | $CH_2Cl \cdot CH_2Cl$ | I | 9·8 | 0·838 | 84 | 13 |
| 1,1,1-Trichloroethane | $CCl_3 \cdot CH_3$ | I | 8·6 | 0·83 | 72–88 | None |

* at 25°C   † at 38°C   ‡ at 15°C   § at 26.5°C   m = minimum   op = open cup

(d)  Flash point.
(e)  Chemical nature.
(f)  Toxicity and smell.
(g)  Cost.

Let us consider these properties in more detail, referring to Table 2, which lists some common solvents together with figures for properties (a)–(d).

### (a)  Solvency

Polymers dissolve by the mechanism described for inorganic solids in Chapter 1. A solvent for a polymer is a liquid, the molecules of which are strongly attracted by the polymer molecules. The molecules of a non-solvent are only weakly attracted. Simple inorganic and organic solids have fixed solublities because, above a certain concentration of molecules in the solution, the strong attractions operate often at short range to overcome the energy of movement of the free solid molecules, so that they group together, take up the crystal pattern and crystallize. The concentration of dissolved solid drops below the limiting figure and crystallization ceases. Amorphous polymers, as we have seen, cannot crystallize. Since the forces of attraction between polymer and solvent molecules are as strong as those between the polymer molecules, dissolving the polymer is like mixing a very viscous liquid with a very fluid one. There is no solubility limit; a true solvent is miscible with the polymer in all proportions.

It is not easy to look at the chemical composition of a polymer and predict solvents for it. A general guide is that like dissolves like, but this is limited, sometimes wrong, and gives no indication of what will happen if a liquid mixture is used. There is still no complete explanation of polymer solvency, but this property was rationalized for the paint formulator in 1955 by the American paint chemist, Burrell. He suggested that for every liquid two factors, or parameters, govern the solvency. The first is the *hydrogen-bonding* capacity of the liquid. He classified paint solvents roughly quantitatively into three groups:

  I.  Weakly H-bonded liquids (hydrocarbons, chloro- and nitro-paraffins).
 II.  Moderately H-bonded liquids (ketones, esters, ethers and ether-alcohols).
III.  Strongly H-bonded liquids (alcohols and water).

For each solvent, a second parameter is calculated from the latent heat of evaporation, by use of the equations of thermodynamics, and the numerical value of this parameter is a measure of the attractive force between the liquid molecules. This parameter is called the *solubility parameter*.

A polymer has a solubility parameter range for each group of solvents. Polymer parameters may be calculated, but are usually obtained by

**Table 3** Spectrum of solubility parameters

| Group I | Group II | Group III |
|---|---|---|
| 6.9 Low odour mineral spirits | 7.4 Diethyl ether | 9.5 2-ethyl hexanol |
| 7.4 *n*-heptane | 8.0 Methyl amyl acetate | 10.3 *n*-octanol |
| 8.2 *cyclo*-hexane | 8.5 Butyl acetate | 10.9 *n*-amyl alcohol |
| 8.5 Dipentene | 8.9 DEG monobutyl ether | 11.4 *n*-butanol |
| 8.9 Toluene | 9.3 Dibutyl phthalate | 11.9 *n*-propanol |
| 9.3 Trichloroethylene | 9.8 2-butoxyethanol | 12.7 ethanol |
| 9.5 Tetralin | 10.4 *cyclo*-pentanone | 14.5 methanol |
| 10.0 Nitrobenzene | 10.7 DEG monomethyl ether | |
| 10.7 1-nitropropane | 12.1 2,3-butylene carbonate | |
| 11.1 Nitroethane | 13.3 Propylene carbonate | |
| 11.9 Acetonitrile | 14.7 Ethylene carbonate | |
| 12.7 Nitromethane | | |

experiment. A selection of solvents, uniformly covering the whole spectrum of parameters, is assembled. A typical selection is shown in Table 3. Attempts are made to dissolve the resin at a technically useful 'solids' level (e.g. 20 per cent for nitrocellulose, 50 per cent for an alkyd) in chosen solvents. If the polymer dissolves in any two solvents in a group, it will dissolve in all the solvents of that group with parameters between those of the chosen pair. Thus the object is to establish the extremes of the range in each group, giving three parameter ranges for the polymer. Table 4 gives parameter ranges for some resins.

An untried liquid will dissolve the polymer, if its own parameter falls within the polymer's appropriate parameter range. Roughly speaking, the parameter of a mixture of liquids is the average parameter. For example, a polyester resin would not dissolve in *n*-butanol (11·4, group III) or xylene (8·8, group I). However, a 1/4 butanol/xylene mixture dissolved the resin. The mixture's parameter is roughly calculated thus:

| | |
|---|---|
| butanol contribution | $1 \times 11·4 = 11·4$ |
| xylene contribution | $4 \times 8·8 = 35·2$ |
| | ___ |
| mixture's parameter | $= 46·6 \div 5 = 9·3$ |

The H-bonding capacity of the mixture corresponds approximately to group II. Note that ethyl acetate (9·1, group II) is a good polyester resin solvent. Where many liquids are involved, the approximation may be too great, but the effect of adding a new liquid to a paint may best be assessed by deciding what effect it will have on the H-bonding characteristics of the liquid mixture and whether it will raise or lower the average parameter.

The ideas of Burrell have been extended by others to make their use more quantitatively precise. Unfortunately, these extensions almost invariably make the concept of solubility parameters more difficult to use in practice. The simple initial concept of Burrell outlined above is still a very useful one for the paint chemist and technologist.

### (b) Viscosity

(i) MEASUREMENT. The viscosity of a simple liquid has been explained and defined in Chapter 1. It is the outward evidence of the internal resistance to flow and is measured in units called 'poises'. Viscosity can be measured by any method involving either the flow of the liquid, or the movement of some object in the liquid.

In the first category come the simple flow cups, perhaps the most widely used consistency control devices in the paint industry. In these, a given volume of liquid is timed as it falls through a hole of definite dimensions.

**Table 4** Solubility parameter ranges of some polymers

| Polymer | Chemical type | Parameter range in | | |
|---|---|---|---|---|
| | | Group I | Group II | Group III |
| N/C, RS 25 cps (dry) | Nitrocellulose | 11·1–12·7 | 7·8–14·7 | 14·5 |
| N/C, SS ½ sec (dry) | Nitrocellulose | 11·1–12·7 | 7·8–14·7 | 12·7–14·5 |
| CAB ½ sec | Cellulose acetate butyrate | 11·1–12·7 | 8·5–14·7 | 12·7–14·5 |
| E/C, N–22 | Ethyl cellulose | 8·1–11·2 | 7·4–11·0 | 9·5–14·5 |
| PMM | Polymethyl methacrylate | 8·9–12·7 | 8·5–13·3 | 0 |
| Acryloid B–72 | Acrylic copolymer | 10·6–12·7 | 8·9–13·3 | 0 |
| Vinylite AYAA | Polyvinyl acetate | 8·9–12·7 | 8·5–14·7 | 0 |
| 45% O.L. linseed glycerol phthalate | Alkyd | 7·0–11·9 | 7·4–11·0 | 9·5–11·9 |
| 30% O.L. soya glycerol phthalate | Alkyd | 8·5–12·7 | 8·4–14·7 | 0 |
| Beetle 227–8 (dry) | Urea formaldehyde | 0 | 0 | 9·5–11·4 |
| Uformite MX–61 (dry) | Melamine formaldehyde | 8·5–11·1 | 7·4–11·0 | 9·5–11·4 |
| Epikote 1001 | Epoxy resin | 10·6–11·1 | 8·9–13·3 | 0 |
| Epikote 1004 and 1007 | Epoxy resin | 0 | 8·4–13·3 | 0 |
| Epikote 1009 | Epoxy resin | 0 | 8·4–9·9 | 0 |
| Epikote 1004 DHC ester | Epoxy ester | 8·5–11·1 | 7·8–9·9 | 0 |
| Propylene glycol maleate phthalate | Unsaturated polyester | 9·2–12·7 | 8·0–14·7 | 0 |

*Note:* Some of the figures in this table do not appear in Table 3. This is because the choice of solvents in Table 3 is merely an example. Other solvents can be, and often are, chosen for parameter range measurements.

A viscosity in poises cannot be obtained, but the flow time is a measure of consistency. The consistency for a spraying paint, for example, can be specified as a given flow time in a particular flow cup at an appropriate temperature (usually 25 °C). If the hole is a long capillary, the answer can be converted to poises, but this sort of apparatus is more practical for pure liquids than for paint. An answer in poises is obtained easily if the viscometer contains a paddle or disc, mechanically driven to rotate the liquid in an enclosed space. The force tending to twist the stationary part of the apparatus in contact with the liquid is usually measured and the scale is calibrated in poises. Comparisons between measurements made in different types of instrument can then be made. Another advantage is that either the rate of shear (proportional to the speed of rotation), or the stress (proportional to the driving force) can be varied, depending on the instrument.

The second category of instrument may involve a ball falling through a liquid, or rolling in a tube containing the liquid. The time taken for movement over a given distance is measured, but conversion of the answer to poises is easily done. Another type, more suitable for very viscous liquids, involves the rise of a bubble in a tube containing the liquid. The time of rise in seconds can be converted to poises. Also for viscous liquids, is a method in which the vibration of a rod is damped by the liquid, the effect being related to the viscosity. Results can be obtained in poises.

(ii) FACTS AND THEORY. There are three essential facts about the viscosity of a polymer solution. Let us consider them in turn:

(1) Early in this chapter it was stated that polymers do not have a solubility limit because they cannot crystallize. Nevertheless there *are* forces of attraction operating between the polymer molecules in solution. Though these forces may be weak compared with those operating in, say, a sodium chloride solution, they operate over a very long length of molecule and there are frequent encounters with other long molecules, enabling those forces to come into play. In addition, there is the possibility of simple mechanical tangling, as with pieces of string. Since both factors have been quoted in Chapter 1 as causes of increasing viscosity, it is not surprising that even comparatively low concentrations of polymer can cause considerable 'thickening' of simple liquids. **As the 'solids' of the solution rise** higher, the encounters and entanglements between molecules become more frequent and **so the viscosity rises,** e.g. solution viscosities for RS ½ sec nitrocellulose (Chapter 11): 12%–1 poise, 20%–10 poises, 30%–100 poises. Eventually the solution becomes so viscous, that it cannot be used in paints unless further liquid is added. At higher 'solids' still, the solution almost ceases to flow and

might be mistaken for a solid. So, although there is no limit at which the solution becomes saturated, there *is* a limit at which the solution becomes too viscous to use. No precise figure can be quoted for this, since it depends upon the use.

(2) If we have two solutions of the same polymer, e.g. polymethyl methacrylate, in the same solvent at the same 'solids', the more viscous solution will contain polymer molecules of higher molecular weight. Since we have the same *total weight* of polymer in both solutions, there are fewer polymer molecules present in the high molecular weight polymer solution, *but they are longer*. The increased opportunity for entanglement reinforced by chemical attraction, outweighs the reduced number of molecules. All the attractions between polymer molecules in a useful solution are continuously forming and breaking with the movement of the molecules. If they were permanent, the solution would not flow, since the polymer molecules would form a semi-rigid, reinforcing network in the liquid. However, to double the molecular weight and halve the number of molecules, involves making a number of these temporary associations into permanent chemical bonds. The network is not made rigid, but it is a good deal less flexible, less amenable to being deformed in flow, than before. So **high molecular weight polymers give more viscous solutions,** e.g. 10% solutions of PVAc of molecular weight 15 000–0·025 poises; 73 000–0·12 poises; 160 000–0·88 poises.

(3) If we take a given sample of polymer of fixed molecular weight and dissolve it at the same 'solids' in a variety of 'true solvents', **the viscosities of the solutions will be proportional to the viscosities of the original solvents,** e.g. 12% polystyrene in methyl ethyl ketone (0·004 poises)–0·4 poises; in ethyl benzene (0·007 poises)–1·6 poises; in *o*-dichlorobenzene (0·013 poises)–3·3 poises. This is important, because we can reduce paint viscosity without lowering 'solids' or polymer molecular weight, simply by changing to a less viscous solvent, if a suitable one is available.

'True solvents' dissolve the polymer at all concentrations. Some liquids give solutions at certain concentrations, but further dilution precipitates the polymer. In these quasi-solutions, polymer molecules collect in clusters, raising the effective molecular weight of the polymer and giving an abnormally high viscosity. Solvents in the middle 80 per cent of the polymer's parameter range are almost certainly true solvents.

It is worth noting here that if true solvents in a solution are partly replaced by non-solvents, the viscosity can rise. This happens when the average solubility parameter of the mixture moves to the extreme of the range for the polymer and clustering of polymer molecules occurs. This is the stage prior to precipitation.

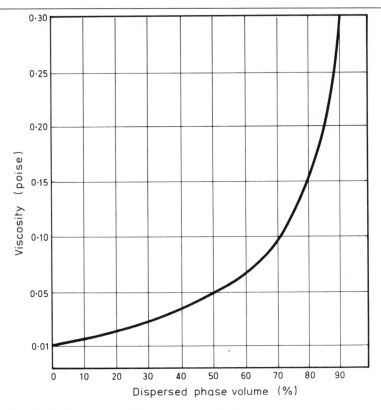

**Fig. 28** Variation of emulsion viscosity with dispersed phase concentration.

Finally, a word about the *viscosity of emulsions*. If a polymer is fully emulsified in a liquid and no part of it is in solution, then the molecular entanglements and associations do not occur. Consequently *the emulsion viscosity must be independent of the molecular weight of the dispersed polymer*. At low solids, the viscosity of the emulsion is the viscosity of the continuous phase liquid. The viscosity of the emulsion rises slightly as the solids 'rise', because of polymer *particle* collisions and weak inter-*particle* attractions, but there is no substantial increase in viscosity, until the concentration of particles becomes so high that the particles can scarcely move at all. A typical solids-viscosity curve is shown in Fig. 28. Thus the paint formulator can use a high molecular weight polymer at high concentrations, *only* by using it in emulsion form.

### (c) Boiling point and evaporation rate

We have already seen that the flow of paint on vertical surfaces can be controlled by solvent evaporation. To get the right solvent balance for doing this satisfactorily, it is necessary to know the relative *evaporation rates* of the solvents available.

In particular we wish to know the evaporation rates of thin films of solvent at normal drying temperatures. These figures are available for many solvents at room temperature (20–25 °C). One system quotes relative evaporation *rates*, with butyl acetate, as reference solvent, rated arbitrarily as 100. *Iso*-propanol at 200 evaporates twice as quickly. Another system quotes relative evaporation *times*. Diethyl ether is reference solvent, with the rating of 1·0. *n*-butanol at 33 takes 33 times as long to evaporate.

In practice, however, we are not concerned with the evaporation of thin films of single liquids under carefully controlled conditions of temperature, humidity and air movement. We are concerned with the evaporation of mixtures of liquids in the presence of polymers under varying atmospheric conditions. Evaporation rates of solvents in mixtures cannot be predicted from the individual evaporation rates: attractive forces between molecules can delay evaporation and these forces will vary from mixture to mixture. Attraction for the polymer molecules will also delay evaporation, so solvents for the dissolved polymer will evaporate more slowly than evaporation rates suggest, while diluents will be unaffected. At best, the published evaporation rates are a rough guide on which to base solvent selection and they are only suitable for paints that lose most of their solvent at room temperature. It is necessary to follow with actual trial of the proposed liquid mixture in the paint. This may need to be repeated several times, with logical replacement of solvents by those evaporating more slowly, to increase flow, or by others evaporating more rapidly, if excess flow (sagging) is occurring.

Because evaporation rates are no more than a guide, many paint chemists feel justified in using the *boiling point* (BP) of a pure solvent, or the boiling range (BR) of an impure solvent or mixture. The boiling point is a fixed property of a liquid and does not vary with the method of measurement, provided that the atmospheric pressure is 760 mm. It is usually true that liquids with low boiling points evaporate more rapidly at room temperature than those with high boiling points. But the boiling point tells you *when* a liquid will *boil*, not *how fast* it will *evaporate* at some lower temperature. When a pair of liquids have boiling points within 30 °C of one another, it is difficult to say which will evaporate more rapidly at room temperature. Alcohols, in particular, are slower than their boiling points suggest, because they are strongly hydrogen-bonded.

In spite of this, boiling points are a sufficiently good guide for most paint purposes. Solvents are graded roughly into three groups: 'low boilers' (BP below 100 °C), 'medium boilers' (BP 100–150 °C) and 'high boilers' (BP above 150 °C). Low boilers are used in spraying paints, because they evaporate between gun and surface and give the necessary rise in 'solids' and viscosity. High boilers are used to give flow and can be the sole solvents where the paint must be kept fluid for fairly long periods, e.g. where brushed paint must 'marry in' when two areas overlap. Medium boilers can be used in all types of paint to give flow at first, followed by fairly quick set-up, but in spraying paints they necessitate greater care by the user, since the nearer the spray-gun gets to the work, the wetter the applied paint will be: the solvent has less time in which to evaporate.

Most paints contain as much diluent as possible, since diluents are usually aliphatic hydrocarbons, which are much cheaper than the true solvents. The limit is decided for the formulator by the fact that the polymer must be kept in solution in the can and at all stages of application and drying. For this reason, the evaporation rates of diluents and solvents have to be balanced carefully, to make sure that there is always enough true solvent present and – particularly – to ensure that the last molecules to evaporate are solvent molecules. If there is too much diluent at any stage, the polymer precipitates and the film appears milky if it is unpigmented, or low in gloss if it is pigmented.

### (d) Flash point

Flash points give an indication of flammability or fire risk. The flash point is the lowest temperature at which enough vapour is given off to form a mixture of air and vapour immediately above the liquid, which can be ignited by a spark or flame under specified conditions. Most countries have regulations concerning the storage, transport and use of products containing the more highly flammable solvents. Highly flammable materials are usually defined as those having a flash point below a certain figure and they may *also* be required to support combustion in a combustion test. The latter clause is introduced so that paints which are safer to use, but have a measurable – and perhaps low – flash point, are not penalized. For example, a waterborne paint may contain a small proportion of low boiling alcohol. This could be detected in a flash point test on the paint, yet the paint might not support combustion. The combination of a flash point below 32 °C and combustibility is used in the *UK Highly Flammable Liquids Regulations*.

Since regulations may vary so much from country to country and, even within one country, may vary according to use or method of transport, it is usually necessary to know accurately the flash point of any paint. The flash point of a mixture of liquids can be lower than the flash point of any of its constituents, so measurements should always be made on the actual sample. These are usually carried out in 'closed' or 'open' cups of varying design and it is important to state which cup was used. Closed cups give lower flash points.

### (e) Chemical nature

It is easy to forget that solvents are chemicals and can react with other paint ingredients. For stability in the can, this is undesirable. Examples of the influence of chemical reactivity on solvent selection are given in Chapters 14 and 15.

### (f) Toxicity and smell

Some liquids, e.g. benzene, have a cumulative poisonous effect and others can be harmful above certain concentrations in the air. The smell of a liquid may be enough to prevent its use for some purposes. Smell is largely a matter of customer's opinion, but toxicity information is obtainable from the sources quoted in the Appendix.

Safe working conditions are established in the UK by the *Occupational Exposure Limits* (OELs) set by the Health and Safety Executive, and in the USA by the *Threshold Limit Values* (TLVs) set by the American Conference of Governmental Industrial Hygienists. OELs and TLVs are concentrations of substances in the working atmosphere which, it is believed, people may be exposed to day after day without adverse effect. They may be quoted as *long-term exposure limits (time-weighted averages* for a normal working day), or as *short-term exposure limits* (maximum concentrations for 10 or 15 min exposure) or absolute *ceiling limits* which should never be exceeded. OELs and TLVs are quoted as parts per million (p.p.m.) or $mg\,m^{-3}$ and are available not only for vapours, but also for dusts. The publications of the H&SE and the ACGIH also draw attention to absorption through the skin where this is an important additional hazard. Examples of long-term exposure limits for common solvents are 50 p.p.m. for *n*-butanol, 100 for xylene, 200 for methyl ethyl ketone, 400 for ethyl acetate and 1000 for ethanol (1986 values). The concentrations of these vapours in the working environment can be measured and, if OELs are being exceeded, it is often possible to solve the problem by changing working methods, e.g. by installing vapour extraction units. Sometimes the alternative of reformulating the paint proves to be the only solution. This is particularly the case when the long-term exposure limit for a solvent is very low (say, less than 10 p.p.m.) and its volatility is high.

### (g) Cost

Hydrocarbons (particularly aliphatic) are the cheapest solvents; some esters, less common ketones and nitroparaffins are the most expensive.

# Ten

# Paint additives

All the main ingredients of a paint shown in the summary on p. 87 have now been discussed in more detail, but one class of ingredient remains: the additives. These are those remarkable materials which, when added to a paint in amounts that can be as little as 0·001 per cent and seldom more than 5 per cent, have a profound influence on the physical and chemical properties of the paint.

## Additives affecting viscosity

In the last chapter we dealt with the viscosities of liquids and polymer solutions. Emphasis was laid on the use of instruments which measure the viscosity in poises. Even better are the types which allow either the stress or the rate of shear to be varied.

With pure liquids, such variations are unnecessary, because the viscosity coefficient is a constant. Referring to Newton's equation in Chapter 1, as the stress increases, the rate of shear does so at the same rate and vice versa. This is also true for polymer *solutions* at paint viscosities. If, however, the 'foreign' material in the liquid is also insoluble in it, the equation may be true no longer. Since most paints contain undissolved materials, it becomes important in many instances to quote not only the viscosity in poises and the temperature of measurement, but also *either* the rate of shear *or* the stress at which the measurement was made.

Why does this deviation from constant viscosity occur? An ideal (pigment) particle dispersion, as we have seen in Chapter 8, consists of a completely uniform distribution of isolated individual particles throughout the liquid. In the same chapter, we saw how chemical groups at pigment surfaces can exert strong attractions over other suitable bodies in the vicinity. Our ideally distributed particles will of course move, since they are affected by gravity and by collisons with fast-moving liquid molecules, or slower but heavier polymer molecules. During the course of these movements they are bound to encounter other particles. Should suitable parts of the surfaces of two particles come into contact, inter-attractions may be strong enough to prevent the particles from separating by their own motion. If this pairing up process continues, quite large groups or clusters may form. A cage-like network or structure may even extend throughout the liquid (see Fig. 29).

Such an arrangement might give the appearance, in the can, of being very viscous, even jelly-like. But once the paint is stirred, the structure breaks up, the particles separate, the liquid 'gives' under the stirrer and the viscosity appears to drop. Should the stirring stop or slacken, the flocculation (as it is called) will begin again and the viscosity will rise. Thus the paint appears to have two viscosities: a high one when it is still and a low one when it is agitated or sheared. In fact, it has a wide range of viscosities, corresponding to all the rates of shear between zero and some value at which the viscosity becomes a minimum, or to all the stages of partial flocculation between complete flocculation and complete deflocculation.

If flocculation occurs slowly, the viscosity, measured at low rates of shear, increases with time during the rest period after an efficient shearing. When this happens the paint is said to be *thixotropic*. If there is no dependence on time or on the previous treatment of the paint and if the viscosity decreases as the rate of shear increases, then the paint is said to be *pseudo-plastic*. If there is a minimum stress required before any flow can occur at all, the viscosity behaviour is said to be *plastic*. All these types of behaviour are, of

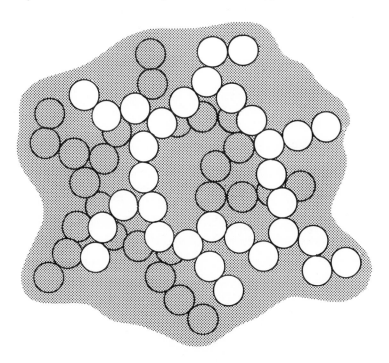

**Fig. 29** A structure of flocculated particles.

course, contrary to Newton's equation and are grouped together under the heading of *non-Newtonian viscosity*.

Most paints show non-Newtonian viscosity to some degree and, if it is marked, it can be most beneficial to the paint. Thus the viscous – even jelly-like – paint in the can does not show hard pigment settlement, applies easily under the shearing action of, for example, a brush or spray-gun, and does not 'sag', because the viscosity rises as soon as the paint is still (or nearly so) on the object being coated. However, non-Newtonian viscosity should be introduced into a paint with caution. While it is true that the paint will not 'sag', it may be extremely difficult to adjust it to obtain reasonable flow-out of application marks as well. Also, if the effect is obtained by pigment flocculation, the perfectly uniform pigment distribution, which gives the maximum light scattering and absorption, will be lost and hiding power will fall. Let us therefore consider ways and means of achieving non-Newtonian viscosity and discuss their relative merits.

### (a) Pigment volume

If the level of pigment in the paint is high, the very bulk of pigment in the paint will cause non-Newtonian viscosity. The close packing of the particles makes some flocculation inevitable and the viscosity will be high simply because the particles impede each other's movements. A further phenomenon, *dilatancy,* can occur at very high pigment volumes. Here disturbing the arrangement of the particles can pack them into an even tighter mass and, on stirring, the viscosity rises and the paint appears to set solid. When stirring ceases, fluidity apparently returns to the paint. Dilatancy, particularly of pigment settlement in cans of paint, is usually highly undesirable.

### (b) Silica and silicates

Dispersion in the paint of a few per cent of very fine particle silica, $SiO_2$ (diameter about $0 \cdot 015 \mu$m), produces a largely pseudo-plastic effect. The surface forces and very large surface area (e.g. $190$–$460$ m$^2$ per gram or $1 \cdot 3$–$3 \cdot 3$ acres per ounce!) are responsible.

Fine particle synthetic aluminium silicates ($Al_2O_3 \cdot 2SiO_2 \cdot 2H_2O$) are also used in paints containing organic solvents. Bentonite is used to impart non-Newtonian behaviour to water paints. It is a complex silicate of the montmorillonite type, containing aluminium, sodium, potassium, iron and magnesium. It has a plate-like particle and will swell to several times its dry volume by adsorption of water between silicate layers.

The metal ions in montmorillonites can be exchanged for quaternary ammonium ions, $R_4N^+$, and when R is a long chain alkyl group, the character of the clay is changed considerably. Organic hydrocarbons are now adsorbed on to the platelet surfaces, because of their attraction for the

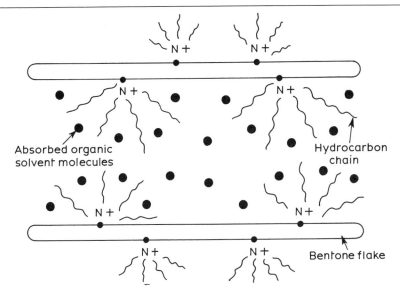

**Fig. 30** The gelling mechanism of 'Bentone'.

alkyl groups. Thus these treated bentonites, given the trade name 'Bentones', will gel organic liquids or impart non-Newtonian viscosity to paints containing organic solvents. Attractive forces between the particles lead to structure throughout the paint (see Fig. 30).

### (c) Resinous thickeners

There are a number of polymers, notably hydrogenated castor oil and its derivatives, polyamides (Chapter 15) and polyamide-oil or polyamide-alkyd reaction products, which can be used to impart non-Newtonian viscosity to paints based on non-polar (mainly aliphatic hydrocarbon) solvents. The thickening mechanisms of polymeric additives are not fully established, but the resins have in common the following features: borderline solubility in the paints in which they are used and chemical structures involving lengthy soluble non-polar chains (e.g. fatty portions of fatty acids) and polar groups, e.g. —OH, —CONH— and —COOH.

The thickening mechanism is probably similar to that of the particles in section (b). A loose network forms, either between very finely divided colloidal particles of resin, via surface forces due to the attractions between polar groups, or between polar portions of large resin molecules. A colloidal particle might have a diameter of $0.01 \mu$m (10 nm) or less. If a single molecule of a polymer of density 1·0 was compactly coiled so as to form a sphere of diameter 10 nm, the molecular weight of the polymer would be 300 000. If a particle of the same size contained solvent molecules, the polymer molecular weight necessary to give that diameter could be much less than 300 000. Depending on the molecular weight of the polymer in question, such colloidal particles in paint could be clusters of 2–100 molecules. The difference between small polymer particles and large molecules is not great.

Resinous thickeners are also used in aqueous paints. The mechanism of operation is similar, but the polarities of the functioning groups are reversed. Thus solubility is conferred by water-soluble polar chain lengths, such as polyethylene glycol or neutralized polyacrylic acid, and attractions form between the water-insoluble (or hydrophobic) portions of the polymer. The latter are usually concentrated as substantial chain segments at the ends of the chains (e.g. as in A–B–A block copolymers), or at intervals along the chains. In emulsion paints (Chapter 11) these *associative thickeners* not only form structures between their own molecules, but also anchor onto the hydrophobic portions of the latex polymer or pigment particles, which thus take part in the structure. High shear viscosity is controlled by thickener concentration, but low shear viscosity is related to the number and strength of the associations that are formed.

We should bear in mind the effect of the solvent environment on the thickeners described so far. In non-aqueous paints, polar groups on particle surfaces are even more accessible to small polar molecules (additives or solvents) than to polar groups on other particles or large molecules. Once these small molecules are in place, they prevent particle–particle or molecule–molecule associations by 'blocking' the active sites. Exactly the same is true (with polarities reversed) in aqueous systems.

Alternatively, the addition of polar solvents to hydrocarbon-based paints will improve the solubility of the thickener molecules and decrease the chance of particles, or of clusters of molecules forming because of border-line solubility. This will lessen or even remove the effectiveness of the thickener. Again the same principles apply to aqueous coatings: additions of water-miscible organic solvents (e.g. ether-alcohols) increase the solvency of the medium for the hydrophobic portions of the associative thickeners and lessen their tendency to associate.

The resinous thickeners above are thermoplastic and all ultimately soluble in some solvent blend or other. As we have seen, their low shear effects can be destroyed by solvent additions. If, however, the polymers are made as latexes or NADs by emulsion polymerization (Chapter 11) and a small amount of cross-linking is introduced via poly-functional monomer, then insoluble colloidal resin particles are produced. These are called *microgels*. In an alternative type of microgel, insolubility is obtained by strong hydrogen-bonding between regular, linear chains, leading to crystallinity, e.g. in polyureas made from hexamethylene diisocyanate and hexamethylene diamine.

Microgels thicken paints by forming flocculated network structures. They are not insensitive to solvent selection, but are much less sensitive and are effective in polar organic solvents. Indeed, some swelling of the polymer particles by solvent probably enhances their effect, perhaps by increasing particle diameter. They can be used in clear coatings without loss of clarity if they are designed to have the same refractive indexes as the film-formers.

## (d) Metal chelates

Metal chelates are organo-metallic compounds in which organic chains are bonded to a central metal atom partly by covalent and partly by co-ordinate bonds, for example:

$$(HO \cdot CH_2 \cdot CH_2)_2 N: \longrightarrow Ti \begin{array}{c} CH_2 \cdot CH_2 \cdot O \quad O \cdot CH(CH_3)_2 \\ \diagdown \quad \diagup \\ \diagup \quad \diagdown \\ (CH_3)_2 CH \cdot O \quad O \cdot CH_2 \cdot CH_2 \end{array} \longleftarrow :N(CH_2 \cdot CH_2 \cdot OH)_2$$

bis (triethanolamine) titanium di-*iso*propoxide

Some chelates of titanium and zirconium have proved especially useful for the thickening of colloid-stabilized aqueous latex paints (Chapter 11). They can induce considerable structure at low shear, leading to non-drip characteristics and good flow with control of runs and sags.

The titanium chelates work best with latexes stabilized by cellulose ether colloids and thickening is thought to occur as the result of hydrogen bonding between hydroxyls in the chelate and in the colloid. An unacceptable irreversible thickening occurs with polyvinyl alcohol (PVA) colloid, due to interchange at the covalent bond between the PVA and the alkoxy group. High surfactant levels interfere with thickening.

The zirconium chelates work with latexes containing sodium carboxy-methyl cellulose or cellulose ethers or polyvinyl alcohol, but in each case the pH of the paint must be carefully controlled.

Both types of chelate are effective at or below 1% and the structure formed is thixotropic. If stirred, the paint slowly recovers structure, but never quite to the original value and there is some loss of structure on each repetition of stirring. Methods of manufacture of the paints must be closely controlled.

## Choice of additives affecting viscosity

The additives that induce non-Newtonian viscosity in paints should be used as sparingly as possible. The paint formulator will find that it is extremely difficult to make several batches of paint to the same formula and the same viscosity and hence the same application characteristics. Some of the resinous thickeners are themselves difficult to reproduce exactly from batch to batch and the mineral thickeners are difficult to disperse reproducibly. As we have seen, all types are influenced by the ingredients of the paint itself and these too may vary slightly from batch to batch. Such variations may not matter normally, but they can play havoc when coupled with the variations in the additive.

The paint formulator must devise suitable tests, which will inform him of variation in viscosity characteristics and must know how to adjust the paint to give the required characteristics. Tests can include measurements of viscosity at several rates of shear (see p. 120). Points plotted on a viscosity-rate of shear graph must fall within a given area. Alternatively, if the paint forms a gel in the can, the strength of the gel in the container can be measured by inserting a razor blade-shaped paddle on a spindle. The paddle is then slowly rotated by a motor. At first the paint and container turn with it against a spring restraining the magnetic platform on which the can stands, but eventually the gel breaks and, at this point, the twisting force or torque is noted and recorded as the gel strength.

The choice of resinous thickener type will depend largely on the solvent system involved. For non-polar organic solvents, linear polymer thickeners are suitable. For polar organic solvents, microgels can be used. For aqueous systems, linear (associative) thickeners, microgels or (if colloids are present) metal chelates can be used. Mineral thickeners are available for all types of solvent.

Compatibility of the soluble resinous thickener with the main binder needs to be checked and excessive amounts of particulate thickeners can reduce gloss, especially in high gloss paints. Large differences in refractive index between binder and particulate thickeners can cause loss of clarity in clear coatings and should be avoided.

The use of high pigment levels is seldom a *chosen* method of introducing non-Newtonian viscosity. Pigment level is usually dictated by gloss and only matt finishes and undercoats have really high pigment levels.

### Additives affecting surface and interfacial tensions

Apart from the defects introduced by the application method, other troubles occur from time to time, often apparently without rhyme or reason. 'Popping' – the appearance of holes or craters in the film – is usually due to inadequate drying at room temperature before the paint is stoved, to the presence of too much low (or medium) boiling solvent, to reaction in the paint producing a gas, or to a faulty undercoat. It can be caused by contamination of the undercoat. 'Blushing', the whitening of the surface of a clear film or the loss of gloss of a pigmented one, is due to condensation and subsequent emulsification of water in the film. If the water evaporates after the film has set, it leaves fine air bubbles in the film. Refraction, reflection and diffraction lead to a milky appearance (see the example of two incompatible resins, Chapter 6). These and other defects are put right by changes in drying conditions. One defect, however, usually requires additives to overcome it. This is known as 'cissing'.

'Cissing' is the appearance of small, saucer-like depressions in the film's surface. These are caused by particles or droplets of incompatible material, which either land on the film during drying, or are present in the paint itself. The weight of the particle causes it to sink but, since there is no attraction between paint molecules and molecules in the particle's surface, the surface tension of the paint resists the particle's entry into the liquid film. Immediately under the particle, the paint molecules have no alternative but to be in close proximity to the particle. In the zone surrounding the particle, the liquid 'skin' is depressed by the particle's weight, but attractions from within the paint film pull the surface molecules away from the particle. There is said to be a high *interfacial tension* between paint and particle. The net result, as Fig. 31 shows, is a 'ciss' mark.

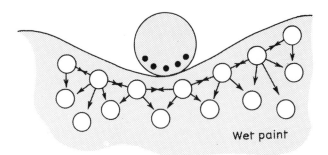

**Fig. 31** A ciss mark.

To prevent cissing, an additive that reduces interfacial tension is required in the paint. When interfacial tension falls, the particle is wetted by the finish and absorbed into the film. Surface-active agents (*Surfactants,* see below) reduce interfacial tension. Alternatively, an agent can be added that will reduce the liquid surface tension so much, that the interfacial tension also becomes low. Silicone oils do this effectively. Very little silicone oil is required, because it finds its way almost entirely to the surface. Silicones are semi-organic compounds of the general type

$$\begin{matrix} R & & R & & R \\ | & & | & & | \\ -Si-O-Si-O-Si- \\ | & & | & & | \\ R & & R & & R \end{matrix}$$

where R may be a hydrogen atom, or some group containing carbon and hydrogen atoms. The higher molecular weight silicones are resins. Since the Si—O and Si—C bonds are much more resistant to heat than C—C bonds, the resins may be used as film-formers in heat resistant finishes. If —R in the general formula is a methoxy (—$O.CH_3$) or hydroxyl group, cross-linking is possible, or reaction with hydroxyl or carboxyl in other resins to produce, for example, *silicone-modified polyester resins*. These are used in Coil Coatings (see p. 152) because of their high resistance to degradation by UV in sunlight. However, the pure resins are very expensive and are incompatible with most of the other paint polymers. They therefore find limited use in paint. As additives, silicone oils are widely used. The silicone oil must be compatible with the finish, or it will itself cause cissing.

Other materials which can prevent cissing are higher molecular weight linear polymers. The mechanism by which they act is obscure.

All such additives are sometimes called 'flow agents'.

**Surfactants**

Surfactants are those chemicals whose molecules have two parts of widely differing polarity and solubility. The soaps, for example, have an ionizing salt 'head' to the molecule and a long non-polar hydrocarbon 'tail'. A different type are the polyoxyethylene ethers of dodecyl alcohol, e.g. $C_{12}H_{25} \cdot O \cdot (CH_2 \cdot CH_2 \cdot O)_6 \cdot H$. These contain the non-polar dodecyl group ($C_{12}H_{25}$—), from dodecyl alcohol, at one end of the molecule and the successive polar ether linkages at the other. Whatever the formula or type, all surfactant molecules have in common these polar and non-polar portions.

Thus one end of the surfactant molecule is attracted to polar molecules (e.g. water) and polar surfaces, while the other prefers a non-polar environment. If we have the problem of two materials which will not wet or make chemical contact with one another, surfactants can bridge the gap. In the example already discussed under 'cissing', we will assume that tiny oily particles are dropping on to the surface of a particularly polar resin solution. There is incompatibility and no wetting. If a suitable surfactant can be found with an aliphatic 'tail' attracted to the particle surface and a polar portion attracted to the resin molecules, chemical contact from particle to paint will be established, interfacial tension will drop and cissing will not occur. The surfactant provides a bridge across the gap between the two unlike molecules.

Wherever this 'bridging' is required, surfactants can be used. If pigment surfaces have poor attraction for binder molecules, surfactants can assist dispersion. When two liquids will not mix, surfactants will stabilize droplets of one liquid in the other, i.e. they will emulsify (Chapter 11). Different surfactants are required for different systems and different applications: the nature and proportions of the two parts of the molecule will vary from use to use.

## Additives affecting gloss

The nature of gloss has been described in Chapter 6. A surface that is sufficiently smooth will be glossy. To reduce gloss we must roughen the surface and break up its smooth outline. In pigmented finishes, this is usually done by increasing the number of pigment particles present in the paint and hence at the surface of the paint. The protruding particles break up the smooth outline. In varnishes or clear wood finishes, we cannot do this and reduction in gloss (to a 'satin' sheen at least) is usually obtained with additives.

Either a few per cent of fine particle silica are used, or else an insoluble wax is dispersed in the finish and floats to the surface during drying. The silica, being an extender, is of course transparent when wetted by the finish. Its high oil absorption makes a small percentage as effective as a much larger quantity of ordinary pigment. However, in many cases the silica, like the wax, tends to float to the surface. It does so as large aggregates containing many particles. The air trapped in the unwetted spaces between the particles reduces the density of the aggregate, allowing it to rise to the surface. When wax is used, only very small quantities are needed and this method gives a very smooth feel to the finish. Polyethylene and polypropylene waxes are particularly efficient and, being relatively high in molecular weight, do not impair the heat resistance of the coating.

## Additives affecting chemical reaction

These are the items described as 'activators' (or 'catalysts'), 'accelerators', 'driers', 'inhibitors' (or 'retarder solutions'). Detailed examples will be given later; for the moment it can be said that they fall into two categories:

(1) Those that initiate the drying reaction (e.g. 'activators')
(2) Those that affect the *rate* of drying (the rest).

An *activator* may be one portion of a two pack paint, one reactant of the resin-forming ingredients. In this instance the activator can hardly be called an additive. The true additive activator is a chemical which, when added as a minor ingredient, sparks off the chemical reaction in the paint. Such an additive is usually a chemical which decomposes to give free radicals which, in turn, initiate an addition polymerization (see *Peroxides,* Chapter 16).

*Driers* (Chapter 12) and *accelerators* are true catalysts in the chemical sense, since they can speed up the chemical reaction responsible for drying, without being consumed in the process. An *inhibitor* (Chapter 16) slows down the chemical reaction, usually by reacting with free radicals, thus preventing them from initiating addition polymerization. Inhibitors are therefore chemically changed in the process. *Retarder solutions* are simply solutions of inhibitor. *Retarder* can also mean a lacquer thinner containing high boiling solvents. These retard drying simply by being slow to evaporate. *Anti-skinning agents* (or *anti-oxidants*) are mild inhibitors of oxidative drying (Chapter 12) and are often included in paints drying by that mechanism, in order to improve stability in the can. They are usually fairly volatile and evaporate once the paint is applied. *Moisture scavengers* are included to inhibit reaction between paint ingredients – e.g. isocyanates or zinc or aluminium pigment – and water, which may have entered the paint during manufacture. Such a reaction might prematurely gel the coating in the container or produce gas under pressure. Calcium oxide has been recommended for zinc-rich coatings and other moisture scavengers are described in Chapter 15.

## Additives affecting living micro-organisms

Paints are susceptible to deterioration brought about by micro-organisms. These include particularly bacteria, yeasts and fungi.

All these types of organism can cause deterioration of liquid paint in the can, but most reported incidents are caused by bacteria. Bacteria can cause gassing, viscosity reduction and colour drift in latex paints (see p. 149). They can enter the paint via infected intermediates and raw materials (including water) or unsterile equipment. The best defence against infection lies in high standards of plant hygiene with regular sterilization of equipment. Nevertheless, the inclusion of an *in-can biocide* at low concentration in latex and paint is a wise precaution. Commercial biocides for paint are often mixtures of complex organic chemicals, which together give protection against a wide range of bacterial types.

Once a paint film has been applied it is exposed to a wide variety of airborne spores from yeasts, fungi and algae. If they find nutrients in the paint or in dirt or debris adhering to it, plus moist conditions ideal for growth, then they will multiply forming unsightly colonies. Algae also require light for growth and so are found mainly on outdoor paint. As the mildew, moulds and algae multiply, film damage eventually results, leading to loss of protective properties. To prevent this *fungicides and algicides* are included in the paint formulation. Again a range of complex organic compounds are effective, as are various organometallic compounds (especially tin complexes). No one compound is completely effective against all species and widest protection is given by a mixture. Selection is based not only on biocidal effectiveness, but also on solubility, stability in the can and long life in the film.

Algae are also found among the large number of plant and animal organisms that cause fouling of ships' hulls below the water line. These organisms, which also include barnacles, tubeworms, mussels, polyzoa, hydroids, ascidians, sponges and sea anemones, settle on the hull and adhere strongly when the ship is stationary or moving very slowly. Once established, they colonize the surface, causing resistance to movement. Reductions of speed by even half a knot lead to large extra costs, arising from lengthened journeys or higher fuel consumption. To prevent such growths, ships bottoms are coated with anti-fouling paints containing so-called *anti-fouling additives*. These are biocides which leach slowly from the film and are effective when released at only a few micrograms cm$^{-2}$ day$^{-1}$. In the paint, concentrations are usually much higher than normal additive levels. Many 'additives' are in fact pigments, including the traditional copper compounds, cuprous oxide and cuprous thiocyanate. They are used at high concentration, either in a slowly dissolving film former (e.g. rosin), or in an insoluble matrix (e.g. chlorinated rubber). Against enteromorpha seaweed (algae), organometallic compounds (often based on tin) are more effective and are used as non-pigmentary solid solutions in e.g. acrylic binders. Even more ingenious is the direct modification of the acrylic binder by copolymerization (Chapter 11) of a tin acrylate or methacrylate (e.g. a trialkyl tin acrylate). The surface of the polymer is gradually eroded away releasing the tin, so that the coating is kept smooth by a 'self-polishing' action.

# Eleven

# Lacquers, emulsion paints and non-aqueous dispersions

This somewhat curious grouping needs an explanation. Although lacquers and emulsion paints involve completely different technologies, they are grouped together here because the leading examples of these types of paints dry by the same mechanism as defined in Chapter 7: lacquer dry. Non-aqueous dispersions must be dealt with alongside aqueous emulsions, since they are complementary.

## Lacquers

Lacquers find widespread use for general industrial finishing (for paper, textiles, plastics and metal), for finishing motor-cars and for finishing furniture. Most of the advantages of lacquers over cross-linked finishes are described in Chapter 7, but the principal reason for their continued success in industry today lies in their ability to harden quickly at all practical temperatures and particularly where oven heating is not available. They are equally suitable where the object to be coated is deformed at modest stoving temperatures, e.g. wood, thermoplastics. They have a big advantage over the so-called 'cold-curing' cross-linking finishes, in that they are supplied in one pack and present no 'shelf-life' or 'pot-life' problems. They also dry faster than paints drying by an oxidative mechanism.

What is a lacquer? It is a finish, clear or pigmented, which consists primarily of a *hard* linear polymer *in solution*. It dries by simple evaporation of solvents (lacquer dry). It is thus possible to make a lacquer from any soluble linear polymer, such as chlorinated rubber, which is used in

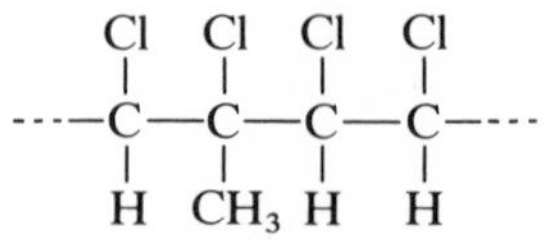

$$----C-C-C-C----$$

with Cl, Cl, Cl, Cl on top and H, CH$_3$, H, H on bottom

chemical-resistant lacquers, but here we have space to discuss only two widely used types: nitrocellulose and acrylic lacquers. Since the nature and properties of a lacquer are largely determined by the main polymer, the chemistry of these two families of polymers will be discussed next.

## Cellulose polymers

These polymers are not wholly man-made (or synthetic), since they are based upon cellulose. This is not made, but found widely in nature, where it forms about half of all the cell wall material of wood and plants. Cotton is almost pure cellulose, and wood pulp is another source.

The cellulose molecule consists of a large number of rings of atoms joined as shown:

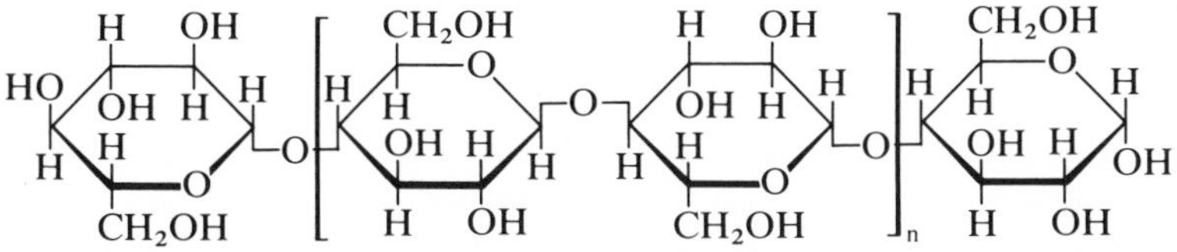

At each corner of each hexagon (except where an oxygen atom is shown) a carbon atom should be imagined. It is omitted here to simplify the diagram.

The repeating unit of the polymer, cellobiose (in the brackets), may be thought of as two glucose molecules linked by the elimination of water

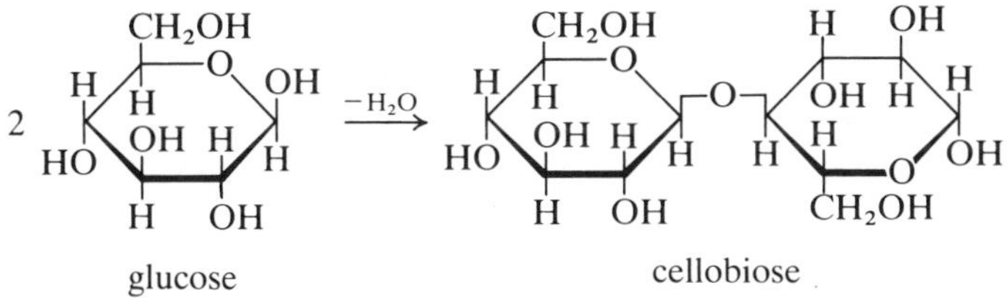

glucose                           cellobiose

Glucose is a sugar, a member of a family of compounds called *carbohydrates*, because their simple formula, $C_x(H_2O)_y$, appears to be composed of *carbon* and *water* (hydrate).

The large number of hydroxyl groups and ether linkages make cellulose sensitive to water, but the molecules are of such immense size (molecular weight 300 000–500 000) and are held together by hydrogen bonding along their lengths to such an extent, that they are not dissolved by water or other normal solvents and can show such uniformity of arrangement of the chains, that the fibres are crystalline in part.

The three hydroxyl groups per glucose ring provide the means of converting cellulose to polymers soluble in organic solvents. They may either be esterified by organic or inorganic acids, or etherified with suitable alcohols. In this way a number of polymers useful in lacquers have been produced:

(1) *Ethyl cellulose*, ether or ethyl alcohol.
(2) *Cellulose acetate*, ester of acetic acid.

(3) *Cellulose acetate butyrate* (C.A.B.), ester of acetic and butyric acids.
(4) *Cellulose nitrate* (nitrocellulose, N/C), ester of nitric acid.

Nitrocellulose is, of course, chemically the wrong name for the polymer, since it contains nitrate ($C\!-\!O\cdot NO_2$) groups and not nitro ($C\!\!=\!\!NO_2$ groups). Nitration can be stopped at any stage by dilution with water and is usually taken to a percentage *nitrogen content* of $10\cdot5$–$12\cdot3$ per cent (or an average of $1\cdot8$–$2\cdot4$ nitrate groups per glucose unit). The molecular weight of the product is too high for paint usage, so the molecules are split by hydrolysis with very dilute acids, aided by heat and pressure:

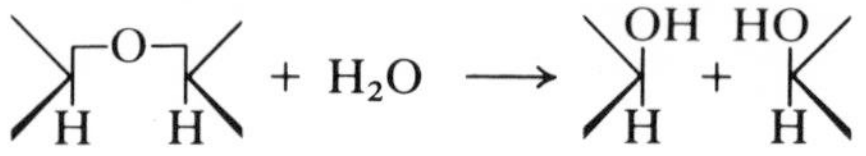

$$\text{(structure)}\ \overset{O}{\underset{H\ \ \ \ H}{\bigg]}\ +\ H_2O\ \longrightarrow\ \overset{OH\ \ HO}{\underset{H\ \ \ \ H}{\bigg]}\ +\ \bigg]$$

Products of molecular weight varying from 50 000 to 300 000 are produced. Water – which would be harmful in a lacquer – is then displaced by an alcohol. N/C must be supplied damped by a liquid; if dry, it is classified as an explosive.

Many grades of N/C, differing in nitrogen content and molecular weight, are available. Nitrogen content controls solubility:

$11\cdot8$–$12\cdot2\%$ N Dissolves in esters, ketones and ether-alcohols.
$11\cdot2$–$11\cdot8\%$ N Dissolves in mixtures of ethanol and esters or toluene.
$10\cdot5$–$11\cdot2\%$ N Dissolves in ethanol.

Molecular weight determines solution viscosity. No international grading system has been agreed upon. Examples of four systems are given in the following table:

| Nitrogen content % | Conc. of N/C in g/100 ml 95% aqueous acetone | Viscosity of this solution in poises at 20 °C | I.C.I. grade (UK) | Hercules grade (USA) | Wolff grade (Germany) | Bergerac grade (France) |
|---|---|---|---|---|---|---|
| $11\cdot8$–$12\cdot2$ | 40 | 8–13 | DHX8–13 | RS ¼ sec | E 400 | E 27 |
| $11\cdot8$–$12\cdot2$ | 40 | 30–50 | DHX30–50 | RS ½ sec | E 560 | E 35 |
| $11\cdot8$–$12\cdot2$ | 20 | 25–45 | DHL25–45 | RS 5–6 sec | E 840 | E 80 |
| $10\cdot5$–$11\cdot2$ | 40 | 8–13 | DLX8–13 | SS ¼ sec | A 500 | A 20 |

**Acrylic polymers**

Acrylic polymers are also a family of polymers, but are entirely synthetic addition polymers. The monomers are mainly esters of the unsaturated acids:

$$CH_2{=}CH-\overset{\overset{\displaystyle O}{\|}}{C}-O-H \quad \text{and} \quad CH_2{=}\overset{\overset{\displaystyle H_3C}{|}}{C}-\overset{\overset{\displaystyle O}{\|}}{C}-O-H$$

acrylic acid        methacrylic acid

Since any alcohol can be used to esterify, there is a wide variety of esters available. Each ester monomer, when polymerized alone, gives a different homopolymer. The number of monomers in a copolymer is not restricted, though some combinations will not copolymerize. The proportion of each monomer may be varied widely. Thus the variety of copolymers possible is boundless.

The acrylate homopolymers are softer and more flexible than the corresponding methacrylates. The latter begin with the hard, tough polymethyl methacrylate (PMM), better known as the plastic with the trade names 'Perspex' and 'Diakon' (UK) and 'Lucite' and 'Plexiglas' (USA). Methacrylate esters of higher alcohols give softer, more flexible polymers and whereas PMM is insoluble in aliphatic hydrocarbons and softens above 125 °C, polybutyl methacrylate is soluble and softens above 33 °C. Poly-*iso*-butyl methacrylate softens above 70 °C, showing the effect of a branched alkyl group. Polylauryl ($C_{12}H_{25}$—) methacrylate is syrupy, but higher alkyl groups lead to hard waxes, which soften at temperatures increasing with the number of carbon atoms in the alkyl group. The non-waxy polymers become more flexible as the size of the alkyl side-group increases, since the polymer chains are kept farther and farther apart, but the waxy polymers are partially crystalline, because the alkyl groups are long enough to align parallel to one another.

Copolymers usually contain a blend of 'hard' and 'soft' monomers, e.g. methyl methacrylate and ethyl acrylate, to give the required properties for the particular purpose. Some of the acid monomer may be included, or the basic amide, e.g. acrylamide, $CH_2{=}CH{\cdot}CO{\cdot}NH_2$. A hydroxyl group may be introduced by copolymerizing with, for example, ethylene glycol monomethacrylate,

$$\overset{\overset{\displaystyle CH_3}{|}}{CH_2{=}C}{\cdot}CO{\cdot}O{\cdot}CH_2{\cdot}CH_2{\cdot}OH$$

### Lacquer film formers

The acrylic polymer which offers the best all-round properties for a metal-coating lacquer exposed to the weather, is the hardest acrylic polymer, polymethyl methacrylate (molecular weight 80 000–150 000). It is hard, clear, scarcely coloured, very little affected by ultra-violet light, insoluble in commercial petrols and resistant to acids and alkalis. Marked deviations from the homopolymer reduce one or more of these properties. It is true that a wide variety of acrylic copolymers are used to make excellent lacquers for many purposes, but in discussing acrylic lacquers we will now confine our attention to lacquers of polymethyl methacrylate.

Polymethyl methacrylate, like nitrocellulose, is a fairly brittle polymer and both must be plasticized for paint uses. It is sometimes convenient to increase flexibility by blending with more flexible resins and other gains – such as improved adhesion – can be obtained. This technique is often used with N/C, though with PMM the film properties are usually down-graded by it and the number of compatible resins that can be used is limited. In fact, it is generally true that the best plasticizers are solvents for the polymer, which have boiling points high enough to ensure little or no evaporation, either at normal atmospheric temperatures, or during moderate stoving schedules. This sort of boiling point (250 °C and above) implies fairly large molecules and liquids more viscous than ordinary solvents, but the substances are simple chemicals, not polymers. The plasticized polymer can be thought of as a high 'solids' solution in plasticizer. We have already seen that less viscous solvents give less viscous solutions. For this reason, less viscous plasticizers give less viscous (or more flexible) plasticized polymer films.

Since group II solvents dissolve most resins, it is not surprising that most plasticizers are esters. Esters of this sort of molecular weight are easily made in wide variety, ketones are not.

The most common plasticizer for N/C is dibutyl phthalate and for PMM, butyl benzyl phthalate:

dibutyl phthalate
B.P. 325 °C

butyl benzyl phthalate
Boiling Range 200–288 °C @ 20 mm
pressure

Compatible resins used with N/C include non-drying alkyds (see Chapter 12), acrylic resins and many natural resins and derivatives, such as dammar and ester gum. Natural resins are mixtures of relatively low molecular weight chemicals, often of uncertain chemistry (see rosin, Chapter 12). Ester gums are produced by reacting polyhydric alcohols, e.g. glycerol, with rosin. An increase in molecular weight occurs, since more than one molecule of monobasic rosin acid will react with one molecule of the alcohol. Replacement of the carboxyl group by an ester linkage increases the range of solvents that may be used. Resins used with PMM include N/C and CAB, other acrylic polymers, and some vinyl polymers. The use of compatible resins with molecular weights in the medium to low range (1000–30 000) can increase substantially the 'solids' of a lacquer at application viscosity.

So far we have seen that our lacquer consists of:

> Pigment (if required)
> The linear polymer
> Plasticizer
> Compatible resin or polymer (if required)

To this we need add only:

> Solvents and
> Additives

and the lacquer is complete. The liquids that are true solvents for N/C and PMM can be seen from Tables 2 and 4, Chapter 9. Non-solvents are also frequently included to reduce cost, but the average solubility parameter of the liquid mixture must lie within the range for the polymer at all stages, from the lacquer in the can to the almost dry film on the surface being coated. Additives depend on the particular requirements of the lacquer. To complete the picture, the formulae of two typical lacquers are given on p. 145. It will be seen readily that they both conform to the same general outline given above.

### Hot spray

One more point about lacquers. Users, who are otherwise satisfied with the properties of lacquers, frequently complain about their low spraying 'solids'. One way of getting higher solids is to apply by hot spray. Here the lacquer is heated on its way to the spray-gun and emerges from the nozzle at 70–90 °C. Since the viscosity of a liquid falls with rising temperature (a result of the increased energy and hence mobility of the molecules), it is possible to start with paint at higher solids and at a viscosity too great for normal spray application. There are two other advantages. Some non-solvents at room temperature become solvents at hot-spray temperatures. Also, cooling between the gun and object provides a viscosity rise, which can prevent sagging. 25–50 per cent higher solids can be obtained by the hot-spray technique.

## Emulsion paints

An emulsion has two parts: the *dispersed phase* (the droplets) and the *continuous phase* (the liquid in which the droplets are dispersed). In Chapter 9 it was explained that the viscosity of an emulsion is little greater than that of the continuous phase. It is not influenced by dispersed phase polymer molecular weight and is influenced by dispersed polymer concentration only at high concentrations. So, if the film-former of a lacquer can be emulsified, then, in principle, the lacquer can be supplied as an emulsion at much higher solids.

## Problems

There are some practical drawbacks to this idea. The first concerns the method of film-formation. When a thin film of polymer *solution* is applied to a surface, evaporation occurs and the polymer molecules are gradually brought closer together as their concentration rises. Ultimately, the polymer molecules are closely packed forming a uniform and *continuous* film. A polymer *emulsion* similarly treated starts to lose continuous phase by evaporation. The polymer droplets (or particles) become more closely packed, until they are touching all their neighbours. At this point we have a *discontinuous* film of polymer, containing some liquid in the voids (or spaces) between the particles. It is now necessary for the particle boundaries to merge, so that ideally the outlines of the particles can no longer be seen, and the film becomes completely continuous.

This merging of particles will not take place unless the polymer molecules in the particles have freedom of movement. Either the glass temperature of the particle must be below room temperature, or heat must be applied. In the former event, this will mean that the final film will be relatively soft, unless the droplets contain a solvent that evaporates later, or unless the polymer can harden by chemical reaction, e.g. by oxidative drying. In practice then, there are three possible types of paint:

(1) A stoving emulsion paint containing a linear polymer.
(2) An emulsion paint containing a linear polymer with a low glass temperature, or one made mobile by plasticizer or solvent.
(3) An emulsion paint containing low molecular weight material that will dry by chemical reaction after film formation, either at room temperature or above.

The tendency to coalescence (as the merging process is called) can be increased by introducing into the paint a high boiling solvent that is miscible with the continuous phase. Some coalescing solvent is left in the voids during the early stages of evaporation, but penetrates the particles during the later stages, giving them greater fluidity. Some time after film-formation is complete, it evaporates altogether.

A second difficulty is pigmentation. Either the pigment must be dispersed in the binder and then the whole emulsified, or it must be dispersed in the continuous phase and mixed with the polymer emulsion later. Often only the latter alternative is available and the dispersion must be carried out with the aid of surfactants. Alternatively it can be achieved using a second resin soluble or dispersible in the continuous phase. Having been dispersed, the

Nitrocellulose glossy wood finish

| | Component | wt. |
|---|---|---|
| **Pigment** | | |
| **Polymer** | N/C DHX 3–5 (or RS⅛ sec.), dry weight | 12·3 |
| **Plasticizer** | Methyl *cyclo*hexanylphthalate | 0·9 |
| | Blown castor oil | 0·9 |
| **Compatible resin** | Ester gum | 13·3 |
| | Hard maleic resin ('Cellolyn' 501) | 8·6 |
| **Solvents** | Butyl acetate | 16·6 |
| | Ethyl acetate | 2·4 |
| | Methyl *cyclo*hexanone | 1·9 |
| | *n*-butanol (N/C damping liquid) | 5·7 |
| | Ethanol | 5·3 |
| | Toluene | 29·6 |
| **Additive to prevent 'bloom'** | 2% Hydroquinone in 2/1 toluene/acetone | 2·5 |
| | | 100·0 |

Ready-for-spraying at 36% solids

---

Grey acrylic motor-car lacquer

| | Component | wt. | |
|---|---|---|---|
| **Pigment** | Rutile titanium dioxide | 11·83 | Roller mill grind stage |
| | Lampblack | 0·12 | |
| | 'Cellosolve' acetate | 2·27 | |
| **Polymer** | 'Paraloid'* A21LV | 12·10 | 2nd and 3rd stages |
| | 'Paraloid' A21LV | 53·32 | |
| **Plasticizer** | Butyl benzyl phthalate | 8·39 | |
| **Solvents** | 'Cellosolve' acetate | 3·47 | |
| | Methyl ethyl ketone | 4·05 | |
| | Toluene | 4·05 | |
| | | 100·00 | |

| *Thinner* | wt. |
|---|---|
| Acetone | 39·5 |
| Toluene | 43·4 |
| 'Cellosolve' acetate | 17·1 |
| | 100·0 |

Thin to 11 secs. No. 4 Ford cup, for spraying at approx. 16% solids.

---

* 'Paraloid' A21LV is an acrylic copolymer, largely polymethyl methacrylate, dissolved at 30% solids in

| | |
|---|---|
| Toluene | 50·0 |
| Methyl ethyl ketone | 40·0 |
| *n*-butanol | 10·0 |
| | 100·0 |

pigment particles must be integrated into the film during coalescence. If a surfactant was used for dispersion, the film-forming resin must flow over the whole pigment surface and wet it during film-formation. If a secondary resin was used, this resin must now merge readily with the main film-former, with which it must be fully compatible. Often, the very mechanism of film formation flocculates well dispersed pigment particles. Failure to integrate the pigment properly can lead to low gloss and film porosity.

The third difficulty concerns application. Emulsions are intrinsically difficult to apply by any method (except, perhaps, spraying), since they have the viscosity of a solvent (the continuous phase). They are too fluid for brushing or dipping, for example. It therefore becomes necessary to raise the viscosity, either with thickener additives, or by use of resinous materials soluble in the continuous phase. The latter convert the continuous phase to a polymer solution, thus increasing its viscosity and therefore that of the paint. In either case, the material added to increase the viscosity must be compatible in the paint, or poor gloss and film integration will result.

### The emulsion

The emulsion can be made in either (*a*) water or (*b*) some other non-solvent. Emulsions based on (*b*) are called organosols or non-aqueous dispersions (NADs). If aqueous emulsions are evaporated to dryness without coalescence of the polymer particles, these may be redispersed in some organic non-solvent to form an *organosol*. However, if the polymer dispersion is produced by direct polymerization of monomers in organic non-solvent, then the product is an NAD. Methods of making emulsions, both aqueous and non-aqueous, will now be considered.

In the chapter on Pigmentation (Chapter 8), under the heading **Dispersion**, we considered how to make stable colloidal dispersions of solid and found that, for stability, it was necessary to keep the particles apart. This could be done by using polymer molecules, anchored strongly to the particle, but also extending out into the solvent, in which they were soluble. These polymer molecules provide a steric barrier around the particle and this method of stabilization is called *steric stabilization*. We also learnt that aqueous pigment dispersions could be stabilized by adsorbed surfactant molecules, which ionized in the water to produce an electrical charge barrier around the particle (*ionic stabilization*). Exactly the same techniques are used to stabilize emulsions.

Emulsions can be made in two ways:

(i) BY MECHANICAL MEANS. The film-former or, if it is too viscous, a solution of the film-former, can be dispersed by vigorous agitation. The polymer may be added to the water, the water to the polymer or both may be added to the mixing vessel simultaneously. The stabilizing surfactant may be introduced mixed with one component or the other. The surfactant may be *anionic*, because in water the active portion is an anion, e.g.

$$R \cdot COONa \xrightleftharpoons R \cdot COO^- + Na^+$$
active portion

or *cationic*, e.g.

$$R_4N \cdot Br \xrightleftharpoons R_4N^+ + Br^-$$

or *non-ionic*, e.g. polyoxyethylene ethers of dodecyl alcohol (see *Surfactants,* Chapter 10). Thus the emulsion particles may be negatively or positively charged, or neutral.

Stability is improved by *protective colloids*, which are polymeric materials with highly polar and non-polar features in their molecular structures. They provide additional steric stabilization. If the protective colloids are not truly compatible with the film-former, gloss will be reduced and the film weakened. Frequently the colloids in water paints (e.g. water-soluble cellulose derivatives) are affected by micro-organisms, so fungicides and bactericides may be included to prevent deterioration.

Emulsions made by mechanical means, using external surfactants and colloids, are not popular in paints today, being formerly used in oil-bound distempers, based on drying oil emulsions. Nowadays, mechanical emulsification is generally used with resins which are internally modified to emulsify in water, e.g. in electrodeposition paints (p. 110).

It is possible to produce condensation polymers in NAD form by mechanically emulsifying the liquid or dispersing the solid ingredients in an aliphatic hydrocarbon of very high boiling point, before carrying out the condensation polymerization. NADs are always sterically stabilized, so the surfactant is polymeric. Frequently *graft copolymers* are used. These contain polar polymer chains, which anchor well onto the dispersed particles and which are chemically bonded (or 'grafted') onto non-polar polymer chains which are soluble in the continuous phase:

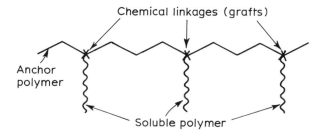

(ii) BY EMULSION POLYMERIZATION. Addition polymerization has been described in Chapter 5. The assumption there was that the monomers were polymerized in bulk or in solution. Aqueous emulsion polymerization begins with a true solution of water-soluble monomers and initiator, or with the 'solubilization' of insoluble monomers and initiator using surfactants. Surfactant molecules can cluster in water, with their fatty organic portions in the centre of the cluster and their water-seeking portions on the outside. Such an arrangement is called a *micelle*. The centre of the micelle is a refuge for insoluble organic monomer molecules, which are attracted by the organic 'tails' of the surfactant molecules. Thus the micelle becomes swollen with monomer, yet the micelle size remains so small that the micelles are not seen by eye and the monomer appears to have been 'solubilized'. Reaction is started by heat, which causes the decomposition of initiator, e.g. peroxide to free radicals. Polymerization to insoluble, emulsifed polymer occurs within some of the micelles, which are 'fed' with monomer by adsorption of other micelles. Further surfactant is attracted to and held at the growing particle surface, stabilizing it. Eventually polymerization goes to completion within the polymer/monomer emulsion droplets.

NAD addition polymers are produced similarly. The continuous phase can be varied quite widely and, by choosing non-solvents of appropriate boiling point, the temperature of the reaction can be controlled by allowing the continuous phase to reflux. The monomers are initially soluble in the continuous phase, but the polymer produced is not. As it precipitates as fine particles, these are stabilized by graft or other copolymers. The particles grow by absorbing further monomer, which polymerizes within the particles.

Most polymers made by emulsion polymerization would be accepted as solids in the dry state. The word 'emulsion' implies a dispersion of one *liquid* in another, so that many people have claimed that these products should not be called polymer emulsions. In practice, then, they are frequently called *polymer dispersions* and the aqueous dispersions are called *latices* (or *latexes*). The paints made from them, however, are usually called emulsion paints.

**Household emulsion paints**

In Europe, the most popular paints of this type are based on vinyl acetate copolymer or acrylic latices. The former are prepared by the emulsion polymerization of vinyl acetate, $CH_3CO.O.CH{=}CH_2$, and are stabilized by a combination of surfactants and protective colloids. Although polyvinyl acetate (PVAc) is a relatively soft polymer, like polymethyl acrylate, its glass temperature is above room temperature. Household paints must coalesce to form a film at room temperature, so PVAc must have its glass temperature lowered, either by addition of a lacquer plasticizer, such as dibutyl phthalate, or more usually by copolymerization with other monomers, e.g. ethyl acrylate, 2-ethyl hexyl acrylate or vinyl versatates.*

* 'Versatic' acids are highly branched carboxylic acids made by Shell. 'Versatic' is a trade name, not a chemical name.

In acrylic latices, the 'hard' monomer is methyl methacrylate and the plasticizing monomer an acrylate, such as butyl acrylate or the two acrylate comonomers mentioned above. Acrylic latices usually contain copolymers of acrylic or methacrylic acid as colloids and thickeners, these being solubilized by neutralisation with base. Whatever the type of latex, coalescing solvents are also added to improve film formation. These may or may not be water-miscible and include alcohols, glycols, ether-alcohols, ether-alcohol esters and even hydrocarbons, all of high boiling point.

Pigment is dispersed in the continuous phase with suitable surfactants, additives and a water-soluble thickener, e.g. hydroxyethyl $(HO \cdot CH_2 \cdot CH_2—)$ cellulose. The dispersion and latex are carefully blended with efficient stirring, to form the paint. It is most stable if just alkaline. The additives include fungicides, to prevent mould growth feeding on the cellulosic or other colloids in the dry film, and biocides, to prevent bacterial degradation of colloid and thickener molecules in the paint can. The bacteria, or enzymes produced by them, rapidly reduce molecular weights of these molecules, leading to a dramatic decline in viscosity and malodorous by-products. Susceptibility to bacterial attack is dependent not only on latex, colloid and thickener chemistry, but also on pH and whether or not other chemicals with a biocidal action, such as certain coalescing solvents or traces of unpolymerized monomer, are present or not.

The paint is easy to apply and dries quickly without unpleasant smells. Brushes and rollers can be cleaned in water. However, it has not been possible, to date, to produce a satisfactory full gloss version. To get a smooth surface of high gloss, fine particle latices must be used and relatively large quantities of polymer in solution. Such paints gives less than perfect flow on application. In addition, it is hard to keep a water-based coating fluid for long enough to give good merging of brushed overlaps and application over porous substrates can lead to low gloss. Acrylic latices and vinylidene chloride $(CH_2{=}C \cdot Cl_2)$ copolymer latices have been used in gloss finishes.

Due to the presence of water-sensitive materials and sometimes (in some matt finishes) very high concentrations of pigment and extender, emulsion paint films retain sensitivity and porosity to water vapour, though this is not as bad as might be feared and many emulsion paints can be washed with safety. Over very porous surfaces, loss of plasticizer or coalescing solvent into the substrate can cause poor coalescence.

For many years styrene-butadiene $(CH_2{=}CH—CH{=}CH_2)$ copolymers were used in the USA in cheap latex coatings for house interiors. More recent developments have been the formulation of primers and undercoats based on latices with good adhesion for wood, the use of ethylene/vinyl chloride/vinyl acetate copolymer latices in house paints and the introduction of 'solid' emulsion paints. *Solid emulsion paint* is in fact paint structured to have a very high low shear viscosity, so that it appears like a stiff jelly in the container. Because of this the paint can be supplied in flat plastic containers which act as ready-for-use roller trays, that can be sealed with a push-on lid after use. Under the high shear of the roller, the structure breaks down, allowing satisfactory application without splashing. Structure of this type is achieved by using higher than normal levels of metal chelate thickener, coupled with latices based on appropriate colloids (see p. 131).

Typical emulsion paint formulae are given opposite.

### Industrial emulsion paints

Based on the principles described in the earlier part of the chapter, various emulsion paints for industrial uses have been formulated and adopted commercially, especially where legal pressures exist to reduce the levels of solvent emitted into the atmosphere from factories. Many of these lacquers dry at room temperature and differ from household paints principally in the formulation of the resin, often an acrylic copolymer, to give a particular combination of properties in the dry coating. Since drying time at room temperature can be considerably prolonged at high relative humidities, force-drying or stoving is often used to achieve consistent production rates. Emulsion primers or undercoats are found more often than higher gloss, high quality topcoats, since flow and levelling are more easily controlled in varying humidity and minor application defects do not spoil the end result. Excellent primers of high corrosion resistance can be formulated on copolymers of vinyl chloride, vinylidene chloride and acrylate esters.

Other industrial emulsion paints are based upon polymers with reactive groups, such as hydroxyl-functional acrylics, reacting with water-soluble amino resins (see Chapter 13) and they are normally stoved. Recently a technique for passivating aluminium flakes in water has been developed and emulsions of this type have been used with aluminium and other pigments to apply the coloured basecoats of basecoat-clear high quality topcoats for motor cars.

An unusual form of industrial emulsion coating, the *autodeposition or autophoretic* process, requires a special mention. In this process, the emulsion particles are *negatively* charged and the coating is maintained at a very low (5–6%) solids content in an acidic continuous phase (pH 2.5–3.5). Extremely clean metal parts are dipped into the coating bath, whereupon acid attack on the metal takes place. From steel, $Fe^{2+}$ ions are produced and are rapidly converted to $Fe^{3+}$ by oxidising agents in the bath. The negatively charged paint particles are destabilized by these multivalent cations, and the destabilized particles flocculate and deposit on the metal surface. Coalescence is deliberately avoided by formulation design, so that a porous film forms, through which acid and cations can diffuse, bringing about more coating deposition. When sufficient coating has been deposited, it is rinsed and dipped in dilute chromic acid (which considerably enhances the corrosion resistance) before it is finally force dried. Because coating forms

## 'Silk' gloss emulsion paint

| | wt % | |
|---|---|---|
| Rutile titanium dioxide | 20·00 | Pigment |
| Calcium carbonate | 7·00 | Extender |
| Sodium polyphosphate | 1·00 | Dispersant |
| Sodium carboxymethyl cellulose | 0·30 | Thickener |
| Proxel CLR | 0·50 | Bactericide |
| Antifoam | 0·30 | |
| Butyl 'Carbitol' acetate | 2·00 | Coalescing solvent |
| Vinyl acetate/vinyl 'Versatate' latex (55% solids) | 42·00 | Film former |
| Ammonia (0·880) | 0·03 | Alkalinity control |
| Water | 26·87 | |
| | 100·00 | |

Ready for brushing at 52% solids.
Proxel CLR is a solution of 1,2-benzisothiazolin-3-one and an amine in water from ICI Organics Division.
Tamol 731 is a surfactant from Rohm & Haas.

## Emulsion-based undercoat

| | wt % |
|---|---|
| Rutile titanium dioxide | 25·0 |
| Calcium carbonate | 2·50 |
| Talc | 12·00 |
| Tamol 731 | 2·50 |
| Hydroxyethyl cellulose | 0·25 |
| Proxel CLR | 0·50 |
| Antifoam | 0·20 |
| Butyl 'Carbitol' acetate | 1·00 |
| Methyl methacrylate/ethyl acrylate/acrylic acid latex (50% solids) | 26·00 |
| Ammonia (0·880) | 0·20 |
| Water | 29·85 |
| | 100·00 |

Ready for brushing at 55% solids.

## Coil coating plastisol

| | wt % | |
|---|---|---|
| Rutile titanium dioxide | 6·00 | Pigment |
| PVC powder | 60·50 | Polymer |
| Dioctyl phthalate | 15·10 | Plasticiser |
| Dioctyl adipate | 4·90 | |
| Epoxy plasticiser | 3·00 | Stabilisers |
| Ba–Cd–Zn stabiliser solution | 1·80 | |
| White spirit | 8·70* | Solvent |
| | 100·00 | |

Ready for reverse roller application at 90% solids. Stove 30–60 seconds to a peak metal temperature of c. 200 °C.
*Or to application viscosity.

wherever metal ions dissolve, it coats all recesses and hidden areas and has good 'throwing power'. Thus it can be considered as doing an electrocoat job without the aid of electricity. Since the composition of the bath is continually changing as it is used, careful analytical monitoring and chemical control is required and this can involve discarding some of the bath contents, as well as the making of additions to the bath.

## Non-aqueous emulsion coatings: organosols and plastisols

Non-aqueous emulsions were referred to on p. 146. If the emulsion is made as an NAD, then a coating can be formulated directly from it, using the principles already described for aqueous emulsion paints, but bearing in mind that dispersants, thickeners and the like must be suitable for a non-aqueous continuous phase. However, there are two types of coating in this overall grouping which are based on polymer made originally by *aqueous* emulsion polymerisation. The emulsions are spray dried and the polymer powder so produced is then redispersed in non-aqueous liquids to produce useful coatings. These coatings are called *organosols* and *plastisols*.

### Organosols

Organosols made from redispersed PVC homopolymer particles have long been used as coatings for the interiors of cans, because of their extensibility and excellent resistance to permeation by the contents of the can. PVC powders are low in cost, but for some years a much more expensive polymer has been used in an organsol to produce high quality coatings for steel and aluminium sheet (coated originally as a *coil* of strip metal, hence the coatings are called *Coil Coatings*). That polymer is polyvinylidene fluoride (PVdF):

$$\cdots-\underset{\underset{F}{|}}{\overset{\overset{F}{|}}{C}}-\underset{\underset{H}{|}}{\overset{\overset{H}{|}}{C}}-\underset{\underset{F}{|}}{\overset{\overset{F}{|}}{C}}-\underset{\underset{H}{|}}{\overset{\overset{H}{|}}{C}}-\cdots$$

It is made by emulsion polymerisation, spray dried and sold as a fine particle powder. Pigment and fluoropolymer particles are separately dispersed in a solution of an acrylic copolymer compatible with PVdF, the two dispersions being blended to give the final paint. After application of paint, the coil is heated to 240–260 °C in 30–60 seconds, when the PVdF particles melt and lose their identity, forming a blend with the acrylic.

PVdF will not absorb UV light from sunlight and so is not degraded by it. If inorganic pigments of equal quality are used with it, the resultant finish, which is also tough and flexible, lasts for 20 years or more outdoors, even in tropical climates, and fully justifies its extra cost. The coated metal sheet is used for the cladding and roofing of buildings.

## Plastisols

A plastisol may be regarded as an organosol in which the continuous phase is almost entirely liquid plasticiser (small amounts of solvent are used for viscosity adjustement). PVC plastisols are made from PVC powder, adipate and phthalate ester plasticisers and minor amounts of epoxy-type resin in solution to aid pigment dispersion and to help (with other additives) to keep the polymer stable to heat and oxidation. The resultant coating is nearly solvent-free and so can be applied in thick films (100–250 microns) and stoved without disruption by escaping solvent. The plasticiser penetrates the particles, aiding sintering as the metal substrate reaches *c.* 200 °C in 30–60 seconds.

These coatings are applied to galvanised steel on a coil coating line. They are very tough and flexible and extremely resistant to building site damage. Hence the sheet is used for cladding and roofing. However, this polymer is subject to slow degradation by UV light and the coatings give 10–15 years life expectancy in temperate climates.

A plastisol coil coating formulation is shown on p. 151.

# Twelve

# Oil and alkyd paints

The paints in this chapter all dry by the second mechanism discussed in Chapter 7 and, in particular, by the same form of that mechanism. This is known as oxidative drying and occurs when the film-former becomes cross-linked as a result of a chemical reaction with oxygen in the atmosphere.

Before proceeding with the chemistry of the oxidative drying process, the reader would be well advised to turn back to p. 41 and re-read the section on oils. He should know the general formula of an oil and, in particular, the arrangement of double bonds in conjugated and non-conjugated fatty acids.

## Oxidative drying

Such are the complexities of the action of oxygen on drying oils that the mechanisms have not been completely unravelled in over three decades of research. The outline that follows is therefore a likely explanation based on the evidence available, but not a completely proven one.

The facts are as follows:

(1) During drying there is a considerable uptake of oxygen. Linseed oil, containing suitable driers (see below), can take up about 12 per cent of its own weight.

(2) Oils containing fatty acids with conjugated double bonds dry much faster than those with non-conjugated unsaturation. For example, without driers and at 25 °C, linseed oil (non-conjugated) requires 120 hours to dry, whilst tung oil (conjugated) requires 48–72 hours. Non-conjugated oils must have a substantial proportion of fatty acids with three double bonds, if they are to dry at room temperature. Fatty acids with two double bonds give good drying at stoving temperatures.

(3) The rate of drying is considerably accelerated by the addition of certain metal soaps (driers), e.g. the same linseed oil dried in 2¾ hours and the same tung oil in 1¼ hours.

(4) In non-conjugated oils, oxygen uptake is accompanied by an increase in hydroperoxide content. (The chemistry of organic peroxides is described in Chapter 16.) Cross-linking and hydroperoxide decompo-

sition occur together and peroxide content falls rapidly until the film has set. In conjugated oils, hydroperoxide formation is not substantial until *after* the film has set.

(5) Decomposition products are produced during the drying process and continue to be produced throughout the life of the film. Drying and further chemical change during the ageing of the film are markedly influenced by ultra-violet light.

## Drying mechanisms

The drying mechanisms must be consistent with the facts. In particular different mechanisms are required for conjugated and non-conjugated oils.

(i) CONJUGATED OILS. Oxygen attacks conjugated double bonds by direct addition, to form free radicals with two unsatisfied valencies (diradicals).

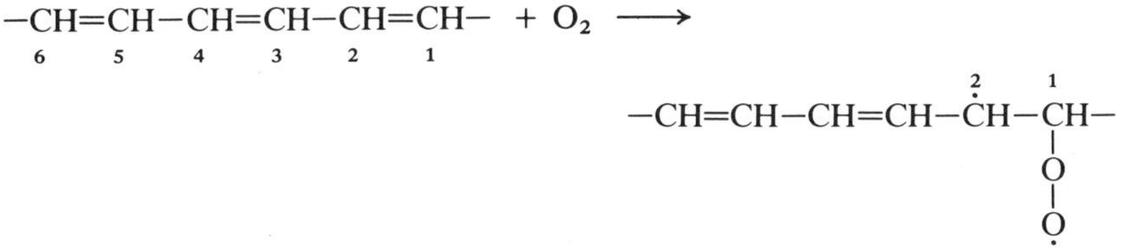

$$-CH=CH-CH=CH-CH=CH- \ + \ O_2 \ \longrightarrow$$

$$-CH=CH-CH=CH-\overset{\bullet}{C}H-CH-$$

Rearrangement of double bonds can take place to give radical sites in the 1,4 and 1,6 positions as well.

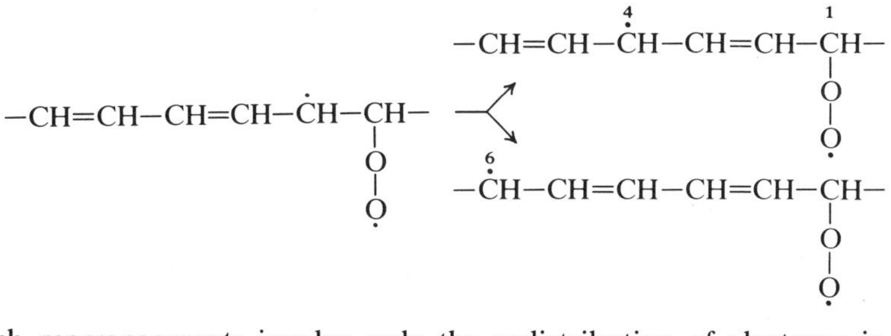

$$-CH=CH-\overset{\bullet}{C}H-CH=CH-CH-$$

$$-CH=CH-CH=CH-\overset{\bullet}{C}H-CH- \quad$$

$$-\overset{\bullet}{C}H-CH=CH-CH=CH-CH-$$

Such rearrangements involve only the redistribution of electrons in the molecule and occur frequently in unsaturated free radicals. Oxygen may react with free carbon valencies to convert them to peroxy radicals.

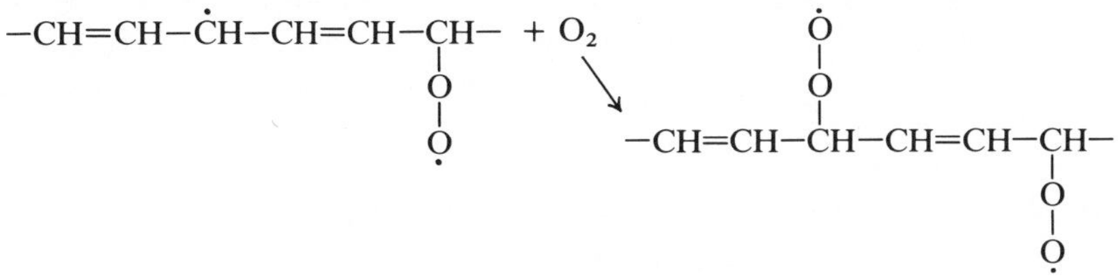

$$-CH=CH-\overset{\bullet}{C}H-CH=CH-CH- \ + \ O_2$$

$$-CH=CH-CH-CH=CH-CH-$$

Cross-linking occurs when the diradicals attack double bonds in other oil molecules. The peroxy radicals produce polyperoxides.

(1)

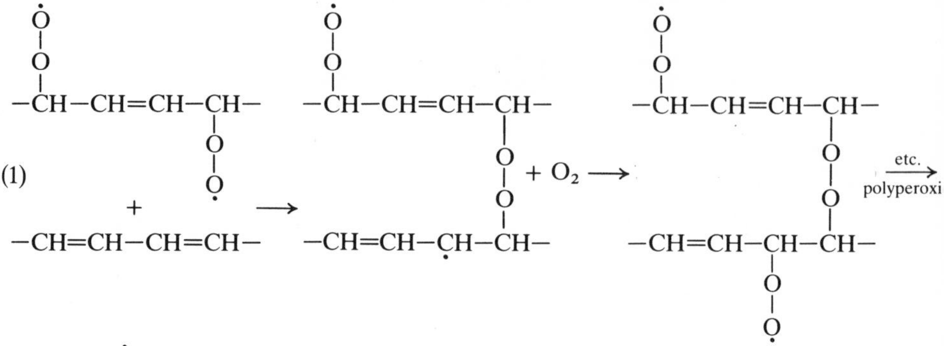

$$
\begin{array}{ccccc}
\overset{\overset{\displaystyle \cdot O}{|}}{\underset{|}{O}} & & \overset{\overset{\displaystyle \cdot O}{|}}{\underset{|}{O}} & & \overset{\overset{\displaystyle \cdot O}{|}}{\underset{|}{O}} \\
-CH-CH=CH-CH- & & -CH-CH=CH-CH- & & -CH-CH=CH-CH- \\
\end{array}
$$

etc.
polyperoxi

(2)

$$
\begin{array}{c}
\overset{\overset{\displaystyle \cdot O}{|}}{\underset{|}{O}} \\
-CH-CH=CH-\underset{\cdot}{CH}-CH=CH- \\
+ \\
-CH=CH-CH=CH-CH=CH-
\end{array}
\longrightarrow
$$

$$
\begin{array}{c}
\overset{\overset{\displaystyle \cdot O}{|}}{\underset{|}{O}} \\
-CH-CH=CH-CH-CH=CH- \\
| \\
-\underset{\cdot}{CH}-CH=CH-CH-CH=CH-
\end{array}
\xrightarrow{\text{etc.}}
\begin{array}{l}
\text{carbon–carbon} \\
\text{cross-linked} \\
\text{polymer}
\end{array}
$$

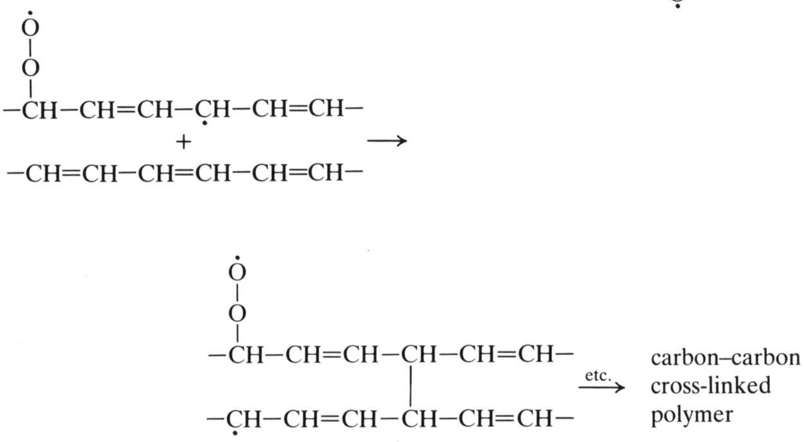

The cross-linking polymerization ceases when the two ends of the growing diradical combine with one another. Polyperoxides are exceptionally stable peroxides, but are decomposed by heat and light to alkoxy radicals (RC—O·), which will in turn react with double bonds to form ether linkages (RC—O—CR).

If 1,4 diradicals are formed from triple unsaturation, the remaining double bonds are non-conjugated and hydroperoxide formation can occur in the manner shown in the next section.

If 1,2 or 1,6 diradicals are formed, the remaining double bonds are conjugated and one will react with oxygen. Since each reaction produces two free valencies, the maximum functionality of eleostearic acid is four and that of the triglyceride twelve!

(ii) NON-CONJUGATED OILS. The essential difference is that reactions between oxygen (or free radicals) and non-conjugated oils occur *without loss of unsaturation.* One school of thought states that oxygen does attack the double bond, but that hydroperoxide formation and rearrangement of bonds occur simultaneously, so that no unsaturation is lost.

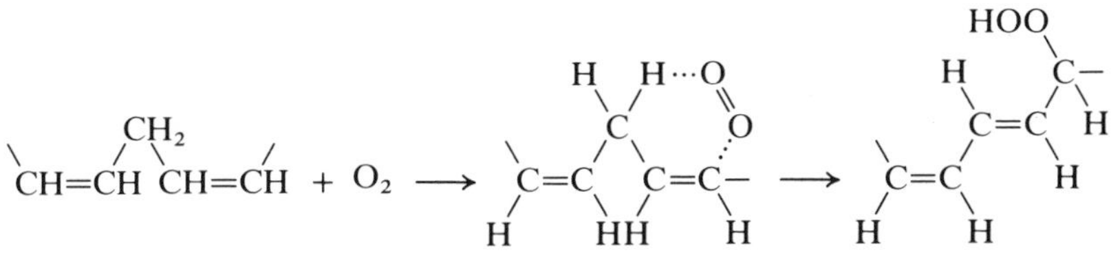

The alternative opinion is that the methylene group between two double bonds is particularly prone to direct oxidation. With oxygen, a hydroperoxide is formed.

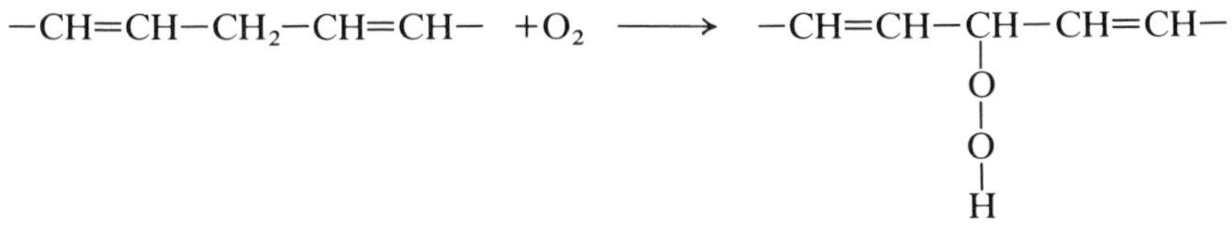

The decomposition of hydroperoxides by driers is described in the section on driers. Free radicals are also obtained when hydroperoxides are decomposed by heat and light:

$$RC—O—O—H \longrightarrow RC—O\cdot + \cdot O—H$$

A free radical will readily remove a hydrogen atom from the active methylene group between double bonds, transferring its free radical nature to the attacked carbon atom. The new free radical may then rearrange to a conjugated form.

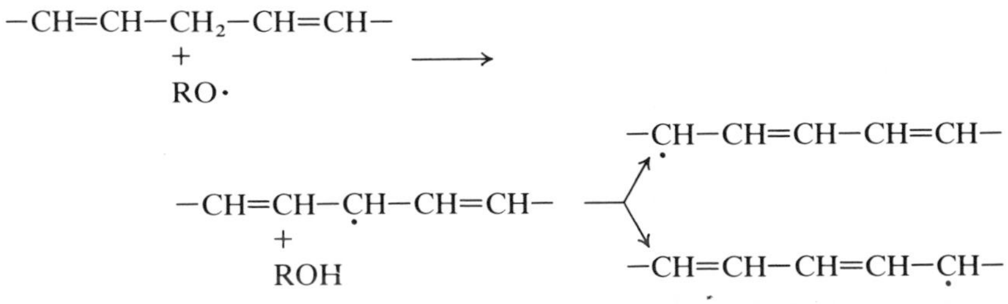

Oxygen can convert these free radicals to peroxy radicals (see 'Conjugated Oils'), which in turn may abstract hydrogen from methylene groups to form hydroperoxides.

Whichever route is correct, it is generally agreed that hydroperoxide is formed and a substantial amount of isomerization to conjugated forms occurs.

During the initial stages of drying, the chief method of cross-linking is by direct combination of free radical sites on different oil molecules (see Fig. 32). Peroxy, ether and carbon-carbon links can be formed.

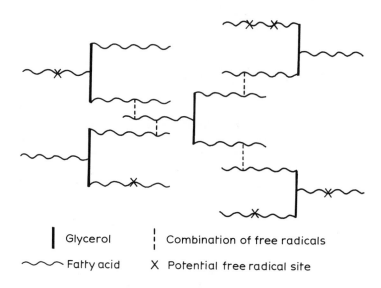

| Glycerol | Combination of free radicals |
| Fatty acid | X  Potential free radical site |

**Fig. 32** The cross-linking of oil molecules.

$$RC{-}O\cdot + \cdot O{-}CR \longrightarrow RC{-}O{-}O{-}CR$$
$$RC{-}O{-}O\cdot + \cdot CR \longrightarrow RC{-}O{-}O{-}CR \Bigg\} \text{ peroxy linkage}$$
$$RC{-}O{-}O\cdot + \cdot O{-}O{-}CR \longrightarrow RC{-}O{-}O{-}CR + O_2$$
$$RC{-}O\cdot + \cdot CR \longrightarrow RC{-}O{-}CR \qquad \text{ether linkage}$$
$$RC\cdot + \cdot CR \longrightarrow RC{-}CR \qquad \text{C—C linkage}$$

Free radicals may also undergo decomposition reactions, e.g.

$$R_1{-}\underset{\underset{\cdot}{O}}{\overset{|}{CH}}{-}R_2 \longrightarrow R_1\cdot + \overset{H}{\underset{O}{\diagdown}}\overset{\diagdown}{C}{-}R_2$$

Aldehydes, ketones and carboxylic acids are known to be decomposition products.

As conjugated structures are formed, they become available for reaction with oxygen and subsequent polymerization in the manner described for conjugated oils. Thus the level of conjugation reaches a peak and then declines, as the conjugated mechanism makes a substantial contribution to drying in the later stages.

(iii) SUMMARY. Conjugated oils dry faster than non-conjugated oils because:

(1) Oxygen attack produces free radicals directly in conjugated oils, but hydroperoxides are the primary products in non-conjugated oils. Free radicals are released by hydroperoxide decomposition.
(2) In the initial stages of drying, conjugated oils yield two free radical sites at every point of oxygen attack, while non-conjugated oils yield only one. For fatty acids of both types, there is one point of attack for every double bond in excess of one. Thus, if the distribution of double and triple unsaturation is similar in both types of oils, conjugated oils have a higher functionality initially.
(3) Although the conjugated mechanism contributes to non-conjugated drying, it does so only in the later stages.
(4) The conjugated oils contain higher proportions of acids with triple unsaturation (see Chapter 3), and therefore have more positions susceptible to oxygen attack.

### Driers

So far the accelerating action of the driers has not been explained. A drier is a metal soap with an acid portion that confers solubility in the oil medium. Synthetic acids, e.g. octoic (or 2-ethyl hexoic, $C_4H_9 \cdot CH(C_2H_5) \cdot COOH$), are normally used. Because of price, the naphthenic acids (derivatives of *cyclo*pentane and *cyclo*hexane) formerly used are now less common. Driers are typical additive materials, being present in quantities generally less than 1%.

Primary driers are true catalysts and contain metals of variable valency, the lower valency being the more stable one, yet capable of oxidation to the higher valency by the products of the drying process. Cobalt and manganese are the principal driers of this type. The metals act in two ways:

(1) They catalyse the uptake of oxygen.
(2) They catalyse the decomposition of peroxide to free radicals.

While cobalt and manganese carry out both functions, lead only catalyses oxygen uptake. Lead is a *secondary* drier, one of a group of metals, including zirconium, calcium and cerium, which assist the drying of the lower layers of the paint film by mechanisms which are not fully understood, but probably involve interactions of the metal with carboxyl (and perhaps hydroxyl) groups in the film-former. Because lead is a cumulative poison, permissable levels of lead in household paints have been reduced to 0.25% (on wet paint) in the EEC and 0.06% (on dry film) in the USA and many paint manufacturers are moving towards drying paints of this type without the use of lead at all.

How driers catalyse oxygen uptake is uncertain, but the decomposition of hydroperoxide by primary driers occurs as follows:

(a)      $Co^{++} + R—O—O—H \longrightarrow Co^{+++} + RO\cdot + OH^-$  ⎫  oxidation
          (cobaltous)                      (cobaltic)                          ⎬  of cobalt
                      or $\longrightarrow Co^{+++} + RO^- + \cdot OH$           ⎭

(b)      $Co^{+++} + R—O—O—H \longrightarrow Co^{++} + RO_2\cdot + H^+$  ⎫
          $Co^{+++} + OH^- \longrightarrow Co^{++} + \cdot OH$                 ⎬  reduction
          $Co^{+++} + RO^- \longrightarrow Co^{++} + RO\cdot$                  ⎭  of cobalt

(c)                          overall
          $2R—O—O—H \longrightarrow RO_2\cdot + RO\cdot + H_2O$

The cobalt is oxidized and reduced over and over again.

Cobalt leads to more rapid drying at the surface of the film than in the lower layers. If the surface dries first, it will be caused to buckle or 'shrivel' when the rest of the film contracts as it hardens. For this reason a 'through' drier, such as zirconium or calcium, is usually added with cobalt to accelerate the drying of the bulk of the film. When the two driers are 'balanced', shrivelling does not occur. Manganese is less pronounced in its surface bias, but is used with secondary driers.

In stoving paints, it is usually satisfactory to include low levels (0.01–0.05% metal on binder) of a single primary drier. Cobalt, manganese, cerium and iron (which only acts as a drier at higher temperatures) are used. Unacceptable discolouration in pale colours can be a hazard, particularly with iron and cerium.

In the late 1970s, a new group of through driers, the aluminium alkoxide derivatives, were introduced. These compounds may have the general formula

$$\begin{array}{c} X \\ | \\ Al—OR \\ | \\ Y \end{array}$$

in which the substituent groups X— and Y— and —OR can react with hydrogen atoms, particularly in carboxyl and (less readily) in hydroxyl groups of alkyds, to bond several alkyd molecules to one atom of aluminium and eliminate volatile by-products from the drying film (HX, HY and ROH). Thus cross-linking through the film is achieved. Careful formulation (particularly adjustment of the carboxyl:aluminium ratio to 1·0 or less) is required to ensure stability in the can and commercial introduction has been confined to high solids coatings based on low molecular weight alkyds. In thick films of these coatings, the improved through dry that these driers can bring has been especially beneficial. Films are claimed to have improved water resistance, gloss retention and yellowing resistance.

## Bodied oils

In Chapter 7, the faster rate of dry to be obtained by beginning with large molecules requiring fewer cross-links was discussed. To speed up the drying of oils, it is logical to polymerize a few oil molecules together before making the paint, stopping the polymerization process before the viscosity of the oil rises excessively. This is called 'bodying' the oil.

One way of doing this, is to carry out the oxidative drying in the bulk oil by bubbling air through it at 75–120 °C. The product is called *blown oil*. Alternatively, driers may be formed in the oil, by heating it to 230–270 °C with metal oxides or salts, which react with fatty acids from the oil. The addition of soluble driers can be carried out at a lower temperature (95–120 °C), accompanied by 'blowing' with air. These products are called *boiled oils* (although in fact the oils do not reach boiling point). The name 'boiled oil' is sometimes given to mixtures of oil and driers blended in the cold, even though no heating has occurred and the oil molecular weight is not in any way increased.

A rather different approach is used to make *stand oils* or *heat-bodied oils*. Here oxygen is excluded. Conjugated oils are bodied rapidly at 200–260 °C. Whereas blown and boiled oils contain principally carbon-oxygen-carbon links between molecules, stand oils contain mainly carbon-carbon linkages. In conjugated oils, these are formed by the reaction known as the Diels–Alder reaction:

$$
\begin{array}{ccc}
\overset{|}{HC}\!\!\diagup^{CH} & \overset{|}{\underset{|}{CH}} & \\
& \overset{|}{\underset{|}{CH}} & \longrightarrow \\
\overset{|}{HC}\!\!\diagdown_{CH} & \overset{|}{\underset{|}{CH}} & \\
& \overset{|}{\underset{|}{CH}} & \\
& \overset{|}{\underset{|}{CH_2}} &
\end{array}
\qquad
\begin{array}{ccc}
& \overset{|}{CH} & \\
HC\!\!\diagup & \diagup^{CH} & \\
\| & & | \\
HC\!\!\diagdown & \diagdown_{CH} & \\
& CH & CH\!=\!CH\!-\!CH_2\!- \\
& | &
\end{array}
$$

For this reaction it is necessary that one molecule should contain two *conjugated* double bonds and the other molecule a double bond. Isomerization to conjugated form, followed by Diels–Alder reaction, accounts for part of the thermal polymerization of non-conjugated oils, but the dominant reaction is thought to arise from the transfer of a hydrogen atom from one molecule to another:

$$
\begin{array}{ccc}
-CH\!=\!CH\!-\!CH_2\!-\!CH\!=\!CH- & & -CH\!=\!CH\!-\!\overset{\bullet}{CH}\!-\!CH\!=\!CH- \\
+ & \longrightarrow & + \qquad\qquad \longrightarrow \\
-CH\!=\!CH\!-\!CH_2\!-\!CH\!=\!CH- & & -\overset{\bullet}{CH}\!-\!CH_2\!-\!CH_2\!-\!CH\!=\!CH-
\end{array}
$$

$$
\begin{array}{c}
-CH\!=\!CH\!-\!CH\!-\!CH\!=\!CH- \\
| \\
-CH\!-\!CH_2\!-\!CH_2\!-\!CH\!=\!CH-
\end{array}
$$

Here only one double bond is lost.

Since the weaker ether and peroxide links are excluded, stand oils are more durable and lighter in colour. The 'body' gives fuller films with higher gloss.

## Oil paints and varnishes

Paints containing oil as sole film-former are scarcely ever used as finishes because of poor gloss, soft films and inferior water resistance. The oils are usually highly pigmented to minimize these disadvantages and find useful outlets as wood and steel primers. The chief oil used is linseed; oils with no triple unsaturation in their fatty acids are too slow drying. Conjugated oils give wrinkled, frosted films on drying, but are usefully mixed with linseed stand oil to increase rate of dry and water resistance. Nowadays there is a tendency to replace oils in primers by alkyds; these permit lower levels of lead driers, or even their complete omission.

It is interesting to note that oil paints were the first 100 per cent solids finishes. The oil is both solvent and film-former and virtually non-volatile. Modern resins of higher molecular weight have led to lower solids, while improving paint performance. Paints combining the present standard of paint properties and very high solids are general targets of research in the paint industry.

Oils are more frequently used in *oleoresinous* paints and varnishes. In these, the oil is either mixed with, or heat-bodied in the presence of, some other resin. The resin may react with the oil to give larger molecules containing fatty acid. Reaction occurs with rosin and its derivatives, phenolic resins and petroleum resins. Other resins, such as terpene resins and coumarone-indene resins, do not react with the oil, but heat helps to dissolve the resin and causes the oil to body. The oils used are those found suitable for oil paints.

An outline of the resins mentioned follows:

*Rosin* consists principally of abietic acid and its isomers.

abietic acid

*Petroleum resins* are produced by the polymerization of the mixture of unsaturated hydrocarbons found in petroleum. As many of these compounds contain more than one double bond, varying amounts of unsaturation can be obtained in the resins.

*Terpene resins* are made by polymerization of terpenes. These lose all unsaturation in the process.

*Coumarone-indene resins* are made by polymerization or copolymerization of coumarone and indene

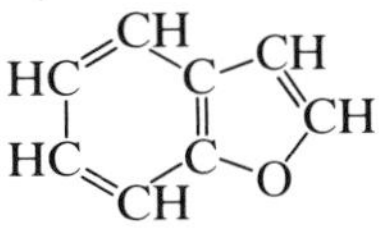

coumarone 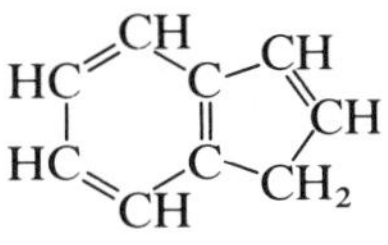                                     indene

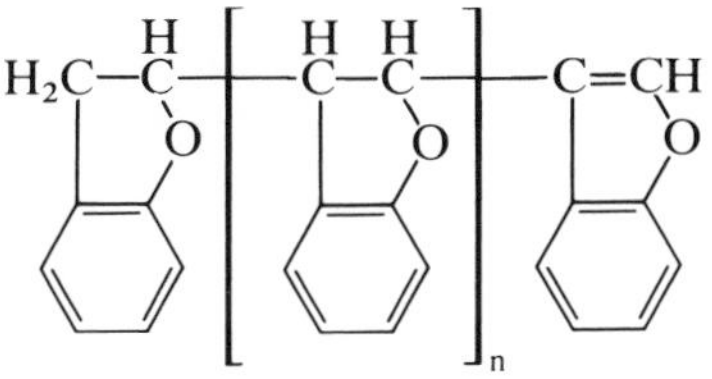

polycoumarone

The resins possess only slight unsaturation, since the attached rings are aromatic.

*Phenolic resins.* A variety of resins is possible, but in each case, the formation of a *phenolic condensate* is the first step:

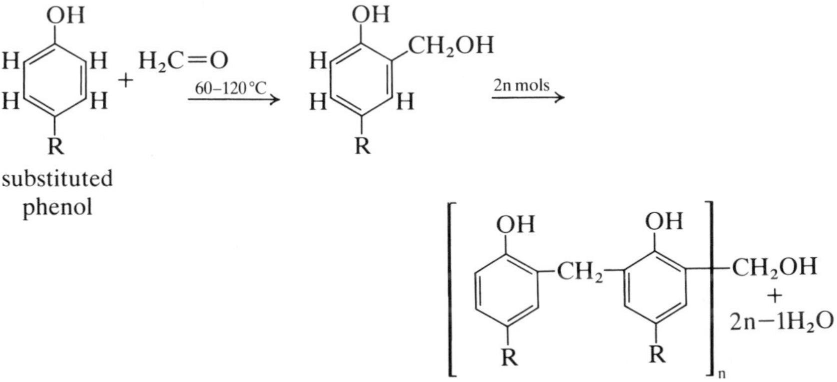

substituted
phenol

$(n = 2 \text{ to } 4)$

With 0·5–1·0 molecules of formaldehyde to one of phenol, an acid catalyst is used. As can be seen, a short-chain linear polymer is produced. The group R— may be hydrogen, an alkyl group or even an aromatic substituent. It need not necessarily be *para*- to the phenolic hydroxyl, but the lightest colour, best solubility in oils and greatest reactivity are obtained with the substituent in this position. This type of phenolic resin is called a *novolac*.

With higher proportions of formaldehyde and the *para*-position vacant, branching is possible. An alkaline catalyst is used to get a slower rate of reaction, thus making it easier to stop the reaction before cross-linking occurs. This type of phenolic resin is called a *resole*.

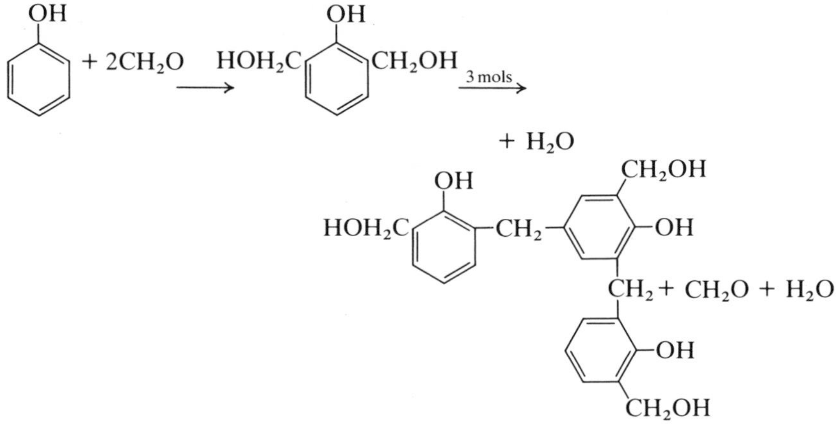

If the *para*-substituent group R— (p. 163) is an alkyl group of about four or more carbon atoms (e.g. tertiary butyl, $(CH_3)_3 \cdot C$—), the product is soluble in oils; otherwise in alcohols, ether-alcohols or alcohol/aromatic hydrocarbon mixtures. If the resin is made from a phenol blocked in the *para*-position, with a high formaldehyde level and using an alkaline catalyst, it will react when heated with unsaturated materials, such as drying oils or rosin to produce a *modified phenolic resin*. The mechanism is thought to be of the type:

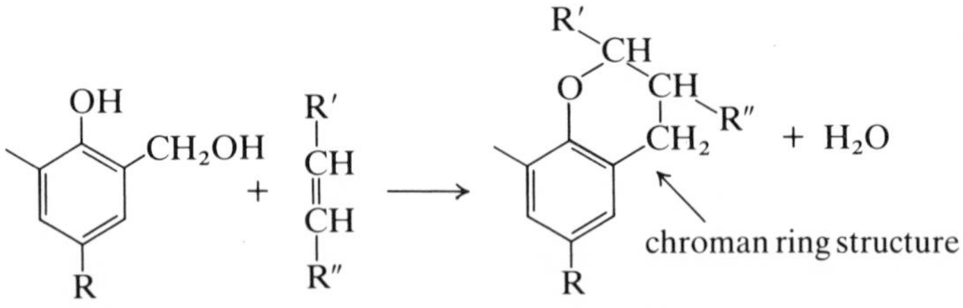

chroman ring structure

The reaction with rosin is used to modify phenolic resins and provides an alternative route to solubility in oils. The rosin carboxyl group is afterwards esterfied with a polyhydric alcohol, such as pentaerythritol, to increase the resin molecular weight even further.

The methylol group in phenolic resins can be butylated, as in nitrogen resins (see Chapter 13) and this improves compatibility with epoxy resins.

The main use for phenolic resins is in oleoresinous varnishes. Apart from the chemical-resistant finishes produced with epoxy resins (Chapter 14), other outlets include modification of alkyds to improve resistance to water and alkali, combination with U/F resins (Chapter 13) to provide metal coatings and combination with polyvinyl formal or butyral resins to produce wire enamels.

The effect of the resin, whatever its type, is to decrease the drying time of the paint, since the resins are hard in their normal states and only require solvent evaporation for drying. It should be noted that the addition of resin means that solvent (white spirit, turpentine or possibly aromatic hydrocarbons) will now be required to reduce the viscosity of the paint, though the low molecular weights of the oils, resins and oleoresinuous products allow medium-high solids at application viscosity.

Other properties are altered by the resins. They harden the film and improve the gloss, though flexibility is reduced. Phenolic, coumarone-indene, petroleum, terpene and other resins free from acid or ester groups, give improved water and chemical resistance. Phenolic resins improve the outdoor durability; unfortunately they give poor initial colour and aggravate the yellowing of the film as it ages. Yellowing is marked when the oil used contains appreciable amounts of linolenic acid (e.g. linseed oil). Manganese driers can also aggravate yellowing.

The weight ratio of oil: resin is called the *oil length*. 3–5 parts of oil to one of resin is described as 'long oil', $1\frac{1}{2}$–3 : 1 as 'medium oil' and $\frac{1}{2}$–$1\frac{1}{2}$ : 1 as 'short oil'. As the oil length increases, the properties contributed by the resin become less and less important. For example, long oil varnishes are the slowest drying, but have the best outdoor durability, because the oil contributes the necessary flexibility.

## Alkyd resins

Alkyd resins provide yet another method of improving the drying potential and film-forming properties of drying oils, yet alkyds need not be made from oils. They are condensation polymers of dibasic acids and dihydric alcohols. The name 'alkyd' comes from its ingredients: *alc*ohols and ac*ids*. An oil-free resin could be made from ethylene glycol and phthalic anhydride:

The polymer is quite linear, as can be seen by following the thickened valency lines. It is a saturated polyester resin, because the polymer chain is held together by a series of ester linkages and it contains no aliphatic double bonds. To include an oil fatty acid we would use a monoglyceride instead of ethylene glycol. This would still leave our alcohol a functionality of two, e.g.:

$$CH_2 \cdot O \cdot CO \cdot (CH_2)_7 \cdot CH{=}CH \cdot CH_2 \cdot CH{=}CH \cdot CH_2 \cdot CH{=}CH \cdot CH_2 \cdot CH_3$$
$$CHOH$$
$$CH_2OH$$

linseed oil monoglyceride (containing linolenic acid). A portion of the alkyd molecule appears like this:

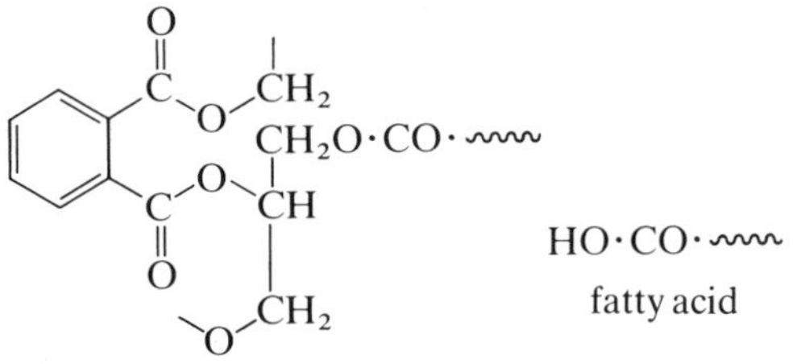

In fact, alkyds are not normally made from monoglycerides, but from oils. The oil is converted to a monoglyceride by heating with glycerol before the phthalic anhydride is added:

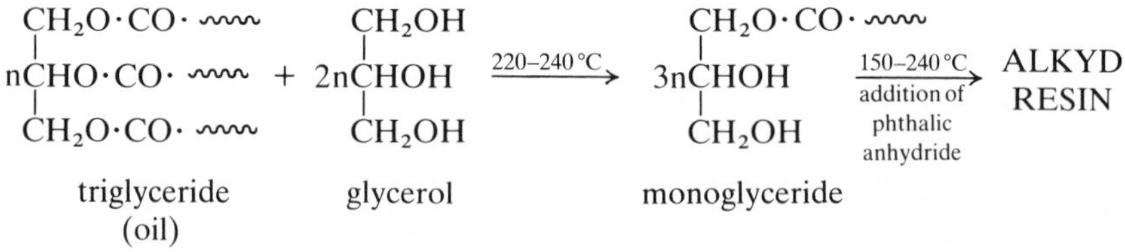

Monoglyceride is formed, because some of the ester linkages in the original oil are broken and the fatty acid molecules produced react with glycerol hydroxyl groups. When this 'redistribution' of fatty acid has gone as far as it can, the *average* molecule will be monoglyceride, though there will be some oil, some glycerol and some diglyceride. Since the oil would body oxidatively in the presence of oxygen, air is displaced from the reaction vessel by inert gas (e.g. nitrogen), which continues to blanket the liquid for the remainder of the process. It is usually the practice to include a solvent for the alkyd (e.g. xylol) as up to 10 per cent of the charge. The solvent accelerates reaction by distilling over into a separator with the water produced. Being immiscible with water, the solvent separates on condensation and is returned to the reaction vessel. Better colour and batch-to-batch reproducibility are obtained.

The course of the reaction can be followed by (*a*) the volume of water produced and (*b*) the drop in acid content, since one —COOH group disappears for every ester link formed. The acid content is measured by titration of resin solution with alcoholic alkali solution and the *Acid Value* (AV) or *Acid Number* of the resin is the number of milligrams of potassium hydroxide required to neutralize 1 g of resin. AVs of commercial alkyds are usually between 5 and 30.

### Ingredients

Many acids and alcohols may be used, but those with more than two functional groups must be kept at a low concentration, so that branching rather than cross-linking occurs, or the groups in excess of two must be neutralized, e.g. hydroxyl with fatty acid. It should be noted that although glycerol has three hydroxyl groups, the secondary hydroxyl group is much less reactive than the two primary groups at 150–180 °C. The secondary group becomes reactive at 200–260 °C. Examples of polybasic acids and polyhydric alcohols used are given below:

*Acids*

phthalic anhydride ⎫
isophthalic acid ⎬ Chapter 4
succinic acid, ⎭
    $HOOC \cdot (CH_2)_2 \cdot COOH$
adipic acid,
    $HOOC \cdot (CH_2)_4 \cdot COOH$
sebacic acid,
    $HOOC \cdot (CH_2)_6 \cdot COOH$
terephthalic acid,

trimellitic anhydride,

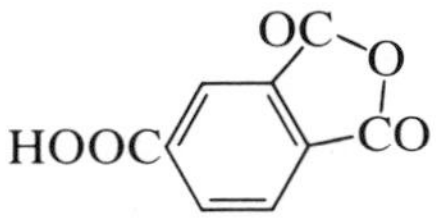

*Alcohols*

Ethylene glycol ⎫
glycerol ⎬ Chapter 3
pentaerythritol ⎭
diethylene glycol      Chapter 9
1,2-propylene glycol,
    $CH_3 \cdot CHOH \cdot CH_2OH$
trimethylol propane,
    $CH_3 \cdot CH_2 \cdot C(CH_2OH)_3$
*neo*-pentyl glycol,
    $HOCH_2 \cdot C(CH_3)_2 \cdot CH_2OH$

Monobasic acids, such as benzoic acid, can be used to stop chain growth and hence to limit molecular weight.

Drying or non-drying oils can be used, according to the type of alkyd required. The non-drying oils give the plasticizing alkyds frequently used in lacquers. The drying oils give drying alkyds. Alkyd molecular weights are of the order of 1000–5000. For a 58 per cent linseed oil glycerol phthalate alkyd (see below), this allows about 2–10 fatty acid chains per average molecule. Thus a drying alkyd will dry faster than the corresponding oil, because of the higher functionality of the molecule and its greater size. Whereas the choice of oil for oil and oleoresinuous paints is limited, alkyds can be made from semi-drying oils, such as soya bean oil, and still have acceptable drying times.

Alkyds, like oleoresinuous varnishes, have an *oil length* and this is expressed as a percentage: the number of grams of oil used to make 100 grams of resin. Below 45 per cent oil is a short oil alkyd, 45–60 per cent medium oil and above 60 per cent, long oil. Long oil alkyds are soluble in aliphatic hydrocarbons, such as white spirit, while short oil alkyds require aromatic hydrocarbons (see Tables 2 and 4, Chapter 9).

Alkyds allow great scope for variation. The polybasic acids and polyhydric alcohols alone can vary the properties, since increasing the number of carbon atoms between the carboxyl groups in the acids (or the hydroxyl groups in the alcohols) leads to more flexible resins. The extent of branching is controlled by acid/alcohol functionality. The alkyd may dry by lacquer dry or oxidatively, according to the oil used. The rate of dry and colour retention will vary with the oil and oil length. The latter controls solubility, cross-linking potential and compatibility with other resins.

The extent and speed of lacquer dry can be increased by modification of the alkyd with polystyrene, poly(vinyl toluene) or methacrylate polymer chains. These may be introduced by carrying out the addition polymerization of the vinyl or acrylic monomer in the presence of (*a*) the completed alkyd or (*b*) the monoglyceride or (*c*) the oil. The reaction with the fatty acid portions proceeds most readily if the double bonds are conjugated. With non-conjugated fatty acids, there is some grafting (see Chapter 11) of the addition polymer onto the active methylene groups, but much of the polymer formed is unattached to the alkyd. Some loss of unsaturation occurs when the copolymerization is successful. This means that the modified alkyd dries more rapidly at first by lacquer dry, but forms fewer cross-links than the unmodified alkyd and so is less solvent resistant and more prone to swelling and wrinkling when recoated with the same paint.

Alkyds are used in stoving and air-drying finishes, alone and with other resins. Their scope is so great that it will be necessary to confine the brief discussion on paints to those containing alkyds alone.

## Alkyd finishes

### Air-drying finishes

Household air-drying finishes are based on long oil alkyds, driers, pigment, white spirit, and additives. Industrial air-drying finishes are frequently based on styrenated or vinyl toluene-modified alkyds. In household paints, linseed oil alkyds are most common, with soya bean oil alkyds used where yellowing is particularly undesirable. Alkyds give better durability out of doors than oleoresinous varnishes. Phthalic anhydride is responsible for their good resistance to degradation by ultra-violet light. The water and alkali resistance of alkyds is not as good as that of the best oleoresinous varnishes, but alkyds have better colour and gloss retention and are less prone to fail eventually by cracking. These are important features in a house paint.

The relatively low molecular weights of alkyd resins allow medium-high solids at brushing viscosity (about 5 poises). Films dry to touch in about 1–4 hours in warm weather, but take longer when it is cold. Once air is allowed into the paint can, drying occurs at the surface forming a skin: it does not usually go deeper. Alkyd films harden slowly over many days. Those designed to last outdoors are soft and easily marked, but very flexible.

### Stoving finishes

Very few stoving finishes contain alkyd alone; chiefly the finish is hardened by blending the alkyds with nitrogen resins (Chapter 13). Short-medium oil length alkyds are used, probably with some proportion of aromatic solvent present.

Examples of coatings based on alkyds, oils and oleo-resinous film-formers are given below.

---

Alkyd gloss finish

|  |  | wt % |
|---|---|---|
| Pigment | Titanium dioxide | 27·0 |
| Resin | 65% O.L. soya bean oil/pentaerythritol alkyd at 75% solids in white spirit | 60·0 |
| Driers | Cobalt octoate (10% Co) | 0·2 |
|  | Zirconium complex (6% Zr) | 0·5 |
|  | Calcium octoate (5% Ca) | 1·7 |
| Solvent | White spirit | 10·6 |
|  |  | 100·0 |

Ready for brushing at 72·5% solids

Oleoresinous varnish for exterior use

| | | wt % |
|---|---|---|
| Resins | Modified drying oil | 16·0 |
| | Oil-soluble, non-heat-reactive phenolic resin | 38·4 |
| Solvents | White spirit | 42·0 |
| | *Iso*-propyl alcohol | 2·4 |
| Driers | Cobalt octoate (6% Co) | 0·3 |
| | Manganese octoate (6% Mn) | 0·1 |
| | Zirconium complex (6% Zr) | 0·8 |
| | | 100·0 |

Ready for brushing at 55% solids

Wood primer

| | | wt % |
|---|---|---|
| Pigment | Rutile titanium dioxide | 16·30 |
| Extender | Magnesium silicate | 34·60 |
| Resins | Long oil soya bean oil/pentaerythritol alkyd at 70% solids in white spirit | 18·20 |
| | Medium oil isophthalic alkyd at 50% solids in VM and P Naptha | 12·10 |
| Additives | | |
| Pigment dispersant | Soya lecithin | 0·25 |
| Anti-settling aid | Bentone 38 | 0·25 |
| Fungicide | Metasol TK100 | 0·10 |
| Anti-skinning aid | Methyl ethyl ketoxime | 0·15 |
| Driers | Cobalt octoate (10% Co) | 0·15 |
| | Zirconium complex (6% Zr) | 0·40 |
| Solvents | Alcohol | 0·10 |
| | White spirit | 17·40 |
| | | 100·00 |

Ready for brushing at 70% solids
*Bentone 38* is a treated clay from N. L. Industries
*Metasol TK100* is 2-(4-thiazolyl) benzimidazole fungicide from Merck

# Thirteen

# Thermosetting alkyd, polyester and acrylic paints based on nitrogen resins

The finishes in this and subsequent chapters dry by the third mechanism described in Chapter 7, that is to say, ingredients *entirely within the paint* undergo a chemical reaction on the surface of the article being coated to produce a cross-linked polymer film. The particular reaction in this chapter is the condensation reaction between methylol groups ($-CH_2OH$) attached directly to a nitrogen atom. For this reason, the family of resins involved are described here as 'nitrogen resins', although they are more widely known as *amino resins*. That name is misleading, since many of the ingredients are amides, not amines.

The principal coatings resins in this group are the urea-formaldehyde (U/F), melamine-formaldehyde (M/F) and the acrylamide- or methacryl-amide-formaldehyde copolymer resins. The reaction concerned normally proceeds rapidly only when assisted by heating or acid catalyst. Thus the finishes in which nitrogen resins are found are principally stoving finishes, though the acid-catalysed cold-curing finishes are particularly important in the woodfinish market. The chief characteristic that nitrogen resins bring to finishes is hardness, although another virtue is their relative freedom from colour. Let us now consider their chemistry.

## Nitrogen resins

What follows must be qualified by saying that the chemistry of the nitrogen resins has not been established beyond doubt. The account presented here is therefore a simplified outline, but one which nevertheless helps the paint and resin formulator to understand the processes involved.

### Urea-formaldehyde resins

U/F resins get their name from their main ingredients: urea and formaldehyde. Both are soluble in water; urea, a white crystalline solid melting at 133 °C, to the extent of 80 g in 100 g of water, while formaldehyde is usually supplied as the 37 per cent aqueous solution, 'formalin', or the solid, paraformaldehyde. 'Paraform' is the condensation homopolymer produced by evaporating formalin:

$$n+_1 CH_2{=}O \ + \ n+_1 H_2O \ \longrightarrow \ \left[ CH_2{\begin{array}{l} {\diagup}OH \\ {\diagdown}OH \end{array}} \right]_{n+1} \longrightarrow$$

$$HO{\cdot}CH_2{\cdot}(OCH_2)_n{\cdot}OH \ + \ nH_2O$$

It yields formaldehyde gas when heated to 180–200 °C. Reactions between urea and formaldehyde proceed as follows:

$$H_2C{\begin{array}{l} {\diagup}NH{\cdot}CO{\cdot}NH_2 \\ {\diagdown}NH{\cdot}CO{\cdot}NH_2 \end{array}} \ + \ H_2O \qquad \begin{array}{l}\text{CONDENSATION} \\ \text{REACTION}\end{array}$$

ACIDIC solution   $+ \ \begin{array}{c} NH_2 \\ | \\ CO \\ | \\ NH_2 \end{array}$   bisamide (insoluble in water) and other insoluble products

$$\begin{array}{c} H{\cdot}N{\cdot}H \ + \ H_2C{=}O \\ | \\ C{=}O \qquad \text{Formaldehyde} \\ | \\ NH_2 \qquad \begin{array}{l}\text{BASIC} \\ \text{solution}\end{array} \\ \text{Urea} \end{array}$$

$$\begin{array}{c} HN{\cdot}CH_2{\cdot}OH \\ | \\ C{=}O \\ | \\ H{\cdot}N{\cdot}H \ + \ CH_2O \\ \text{monomethylol} \\ \text{urea} \end{array} \quad \longrightarrow \quad \begin{array}{c} HN{\cdot}CH_2{\cdot}OH \\ | \\ C{=}O \\ | \\ HN{\cdot}CH_2{\cdot}OH \\ \text{dimethylol urea} \\ \text{(soluble in water)} \end{array} \quad \begin{array}{l}\text{ADDITION} \\ \text{REACTION}\end{array}$$

The methylol group gets its name because it consists of a *methyl* group in which one hydrogen atom is replaced by an alcoho*l*ic hydroxyl. Note particularly that acid conditions give a condensation reaction, while basic conditions give an addition reaction. It is the latter route that interests us at present, because the water-soluble dimethylol urea could at this stage be converted to a water soluble resin, by a switch to acid conditions:

$$HN{\cdot}CH_2OH \ + \ HN{\cdot}CH_2{-} \ \xrightarrow[\quad]{\begin{array}{c}\text{acid} \\ \text{solution}\end{array}}$$

$$HN{\cdot}CH_2{\cdot}N{\cdot}CH_2{-} \ + \ H_2O \qquad \begin{array}{l}\text{Condensation} \\ \text{reaction} \\ \text{(polymerisation)}\end{array}$$

Dimethylol urea has a functionality of four (two —CH$_2$OH, two HN—), so resin formation occurs with ultimate cross-linking. The high concentration of hydroxyl groups present in the linear version of the resin means poor solubility in organic solvents. This situation can be remedied by reaction of the hydroxyl with a suitable alcohol

$$HN \cdot CH_2OH + HO \cdot R \xrightarrow[\text{solution}]{\text{acid}} HN \cdot CH_2 \cdot O \cdot R + H_2O$$

Condensation reaction (etherification)

ether link

If the alcohol R·OH is ethyl alcohol, the product is soluble in ethyl alcohol, but if it is butyl alcohol, then the product is soluble in aromatic hydrocarbons (or their mixtures with alcohols).

Thus to make a resin suitable for non-aqueous paints, urea, formalin and butanol are placed in the reaction vessel and heating is commenced with the mixture slightly alkaline. When sufficient of the two methylol ureas have been formed, the mixture is made acid. The two condensation reactions described above – polymerization and etherification – now proceed in direct competition. The nature of the final resin will depend upon the proportions of the three ingredients, the amount and type of acid and the temperature and time of reaction. The proportions might be 1 mol urea: 2 mols formaldehyde: 1 mol butanol. An idealized reaction scheme would be:

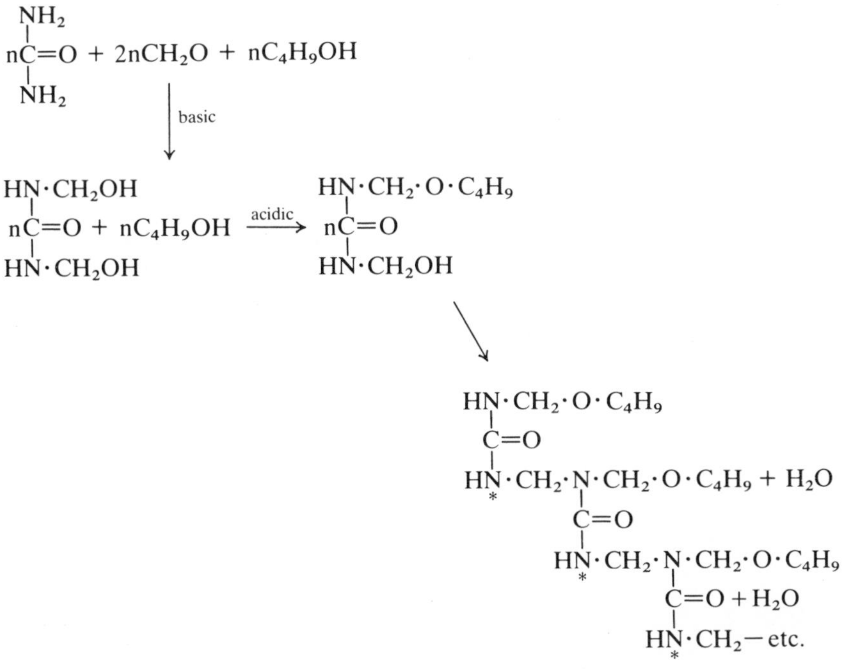

$$\begin{array}{c} NH_2 \\ | \\ nC=O + 2nCH_2O + nC_4H_9OH \\ | \\ NH_2 \end{array}$$

$$\downarrow \text{basic}$$

$$\begin{array}{c} HN \cdot CH_2OH \\ | \\ nC=O + nC_4H_9OH \\ | \\ HN \cdot CH_2OH \end{array} \xrightarrow{\text{acidic}} \begin{array}{c} HN \cdot CH_2 \cdot O \cdot C_4H_9 \\ | \\ nC=O \\ | \\ HN \cdot CH_2OH \end{array}$$

$$\begin{array}{c} HN \cdot CH_2 \cdot O \cdot C_4H_9 \\ | \\ C=O \\ | \\ HN \cdot CH_2 \cdot N \cdot CH_2 \cdot O \cdot C_4H_9 + H_2O \\ \overset{*}{} \quad | \\ C=O \\ | \\ HN \cdot CH_2 \cdot N \cdot CH_2 \cdot O \cdot C_4H_9 \\ \overset{*}{} \quad | \\ C=O + H_2O \\ | \\ HN \cdot CH_2 - \text{etc.} \\ \overset{*}{} \end{array}$$

It is most unlikely that only a linear polymer will be formed, because the reactive HN—* groups in the growing polymer molecule can react with —CH$_2$OH groups in other molecules to produce branching. Rings could be formed:

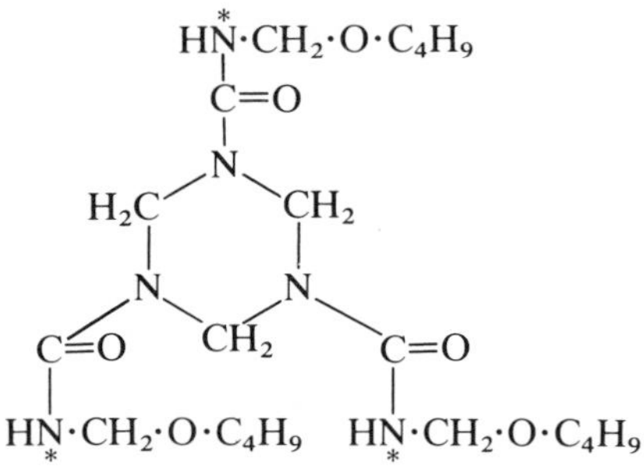

$$
\begin{array}{c}
\overset{*}{HN}\cdot CH_2\cdot O\cdot C_4H_9 \\
| \\
C=O \\
| \\
N \\
H_2C \diagup \quad \diagdown CH_2 \\
| \qquad\qquad | \\
N \qquad\qquad N \\
C\overset{/}{=}O \quad CH_2 \quad C=O \\
| \qquad\qquad\qquad | \\
\underset{*}{HN}\cdot CH_2\cdot O\cdot C_4H_9 \quad \underset{*}{HN}\cdot CH_2\cdot O\cdot C_4H_9
\end{array}
$$

with chains leading from the rings via reaction at *. Consequently, the reaction is stopped by cooling the mixture and neutralizing the acid, while there are still appreciable numbers of HN— and —CH$_2$OH groups unreacted. The resin may later be 'cured' to a cross-linked, insoluble state by allowing the reaction to proceed further on the coated article:

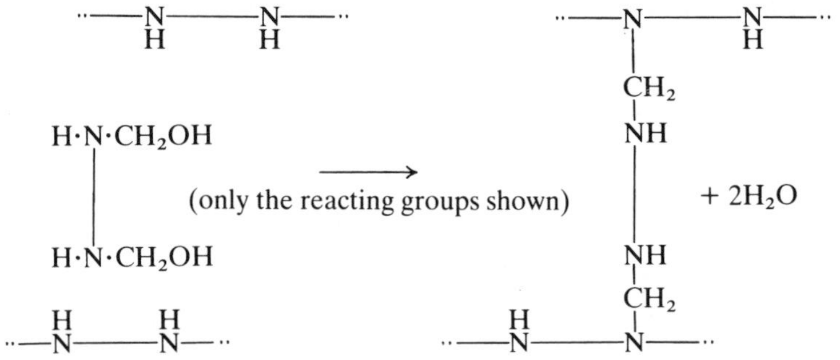

$$
\begin{array}{ccc}
\cdots—N——N——\cdots & & \cdots—N——N—\cdots \\
\quad\,\, H \qquad H & & \qquad | \qquad\quad H \\
& & \qquad CH_2 \\
H\cdot N\cdot CH_2OH & & \qquad NH \\
| & \longrightarrow & \qquad\qquad\qquad\qquad +\,2H_2O\\
| & \text{(only the reacting groups shown)} & \qquad NH \\
H\cdot N\cdot CH_2OH & & \qquad CH_2 \\
\quad H \qquad H & & \qquad H \qquad | \\
\cdots—N——N—\cdots & & \cdots—N——N—\cdots
\end{array}
$$

In all this the butylated hydroxyls are not entirely inactive, since butanol is one of the products of the cross-linking reaction. The reaction may be

$$
\overset{|}{\underset{|}{N}}\cdot CH_2\cdot O\cdot C_4H_9 + H\cdot \overset{|}{\underset{|}{N}} \longrightarrow \overset{|}{\underset{|}{N}}\cdot CH_2\cdot \overset{|}{\underset{|}{N}} + C_4H_9OH
$$

but it occurs less readily than the reactions involving non-etherified hydroxyl, since one of the effects of increasing the extent of butylation is to decrease the curing rate of the resin. This is also reduced by choosing an etherifying alcohol containing more carbon atoms (e.g. nonanol, C$_9$H$_{11}$OH). However, both of these changes serve to increase the solubility of the resin in organic solvents and to improve its compatibility with other resins. A balance between reactivity and solubility is required in coatings resins. *n*- and *iso*-butyl alcohols are most frequently preferred.

## Melamine-formaldehyde resins

Melamine is a white, crystalline solid melting at 354 °C and only slightly soluble in water. It is made as follows:

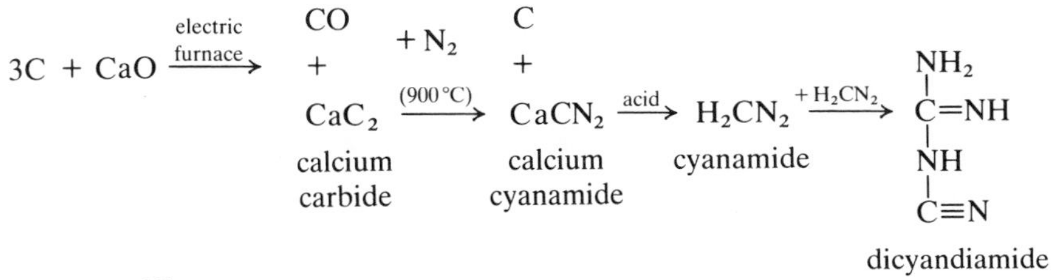

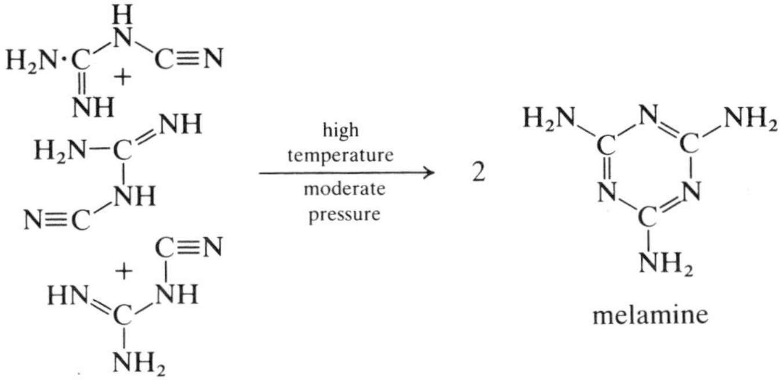

$$3C + CaO \xrightarrow{\text{electric furnace}} \begin{matrix} CO \\ + \\ CaC_2 \\ \text{calcium carbide} \end{matrix} \xrightarrow[\text{(900 °C)}]{+N_2} \begin{matrix} C \\ + \\ CaCN_2 \\ \text{calcium cyanamide} \end{matrix} \xrightarrow{\text{acid}} \underset{\text{cyanamide}}{H_2CN_2} \xrightarrow{+H_2CN_2} \begin{matrix} NH_2 \\ | \\ C{=}NH \\ | \\ NH \\ | \\ C{\equiv}N \end{matrix}$$

dicyandiamide

3 molecules of dicyandiamide → (high temperature, moderate pressure) → 2 melamine

The primary amino groups of melamine react with formaldehyde as do those in urea, but with a difference. *Both* hydrogen atoms may be converted to methylol groups, making a maximum of six for the molecule. If melamine, formaldehyde and butanol are reacted together under acid conditions, a butylated resin can be obtained:

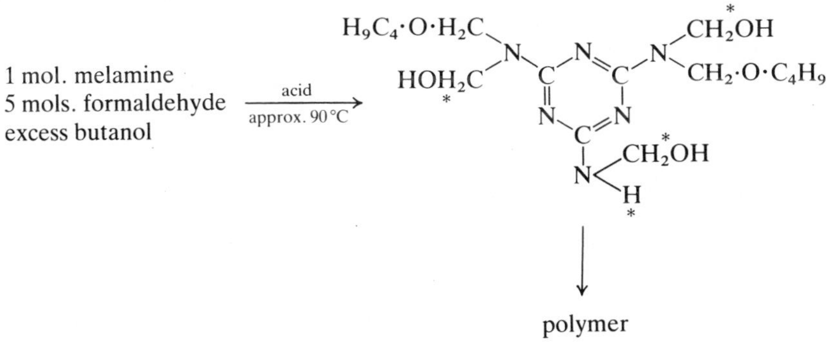

1 mol. melamine
5 mols. formaldehyde $\xrightarrow[\text{approx. 90 °C}]{\text{acid}}$
excess butanol

↓

polymer

The reactive groups marked * undergo the condensation reactions shown for U/F resins, as well as etherification by condensation between methylol groups:

$$-CH_2OH + HOCH_2- \longrightarrow -CH_2 \cdot O \cdot CH_2- + H_2O$$

The polymerization is stopped short of cross-linking which, with further heating, will occur rapidly due to the high functionality.

If melamine is converted to methylol melamine with formaldehyde and then acidified and etherified with a very large excess of methanol, polymerization can be largely avoided and the end product is essentially hexamethoxymethyl melamine (HMMM):

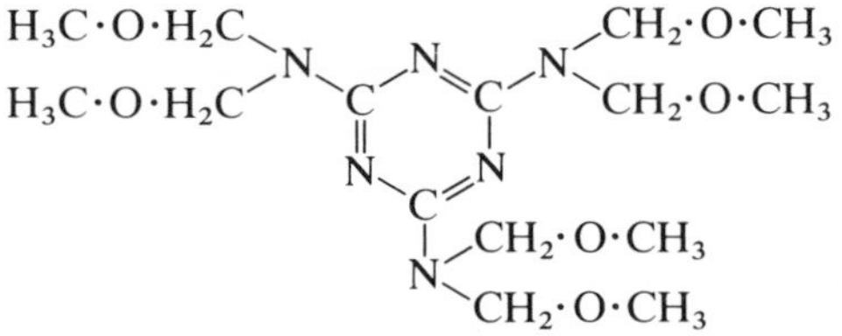

Such HMMM resins generally contain an average of not less than 5·5 methylol groups per molecule of melamine and not less than 5·5 methoxy ether groups. They are soluble in all common organic solvents except aliphatic hydrocarbons and also dissolve in water alone, or mixtures of water and small amounts of an organic solvent (e.g. ethanol). They have very low molecular weights and so can be used in coatings of high solids and low viscosity. Greater reactivity can be obtained by going for less than full methylolation and/or less than full etherification by methanol. Some condensation inevitably occurs, with an increase in molecular weight and viscosity. Such resins are methylated melamines rather than true HMMMs.

Without the addition of acid catalysts or presence of carboxyl groups, HMMM resins require high temperatures for reaction, but with acids present, reaction occurs at temperatures from room temperature to 150–175 °C. The methoxymethyl groups react with hydroxyl, carboxyl and amide groups. HMMM molecules react preferentially with the other resins rather than with each other and, although six groups are theoretically available for reaction, in practice only three normally react.

Thus HMMM resins are very suitable for a wide variety of finishes, but particularly for water-based or high solids content finishes.

### Acrylic nitrogen resins

The two long-established nitrogen resins have the following features in common:

(1) In the monomer, —NH₂ is attached to carbon, this carbon atom being

adjacent to an atom or group with a greater attraction for electrons (e.g. $H_2N-C-$ in urea, where the oxygen is strongly electron attracting

$$\underset{O}{\overset{\|}{}}$$

and $\overset{\overset{/}{N}}{\underset{\underset{NH_2}{/}}{\overset{\|}{\underset{C-N}{}}}}\overset{//}{}$ in melamine, where the nitrogen is electron-attracting).

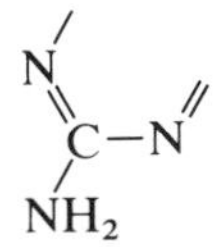

The $-NH_2$ group is capable of methylolation and the methylol groups are capable of etherification.

(2) The monomer must have a functionality of 3 or more if the resin produced is to be cross-linked ultimately.

Such features have been assembled in other monomers to give useful nitrogen resins. Cost considerations have prevented most of these from commanding wide usage. The same features and cross-linking mechanism can be built into acrylic polymers. The polymer may be a copolymer of any suitable acrylic or vinyl monomers, but must contain copolymerized amide (e.g. acrylamide). The amide group is thus attached to the polymer chain.

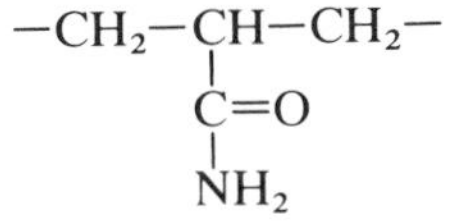

$$-CH_2-\underset{\underset{NH_2}{|}}{\underset{C=O}{|}}{CH}-CH_2-$$

and bears a marked resemblance to urea.

The $-NH_2$ group can be methylolated with formaldehyde and the methylol group etherified (e.g. with butanol). Sufficient amide is copolymerized to give the average polymer molecule a functionality well in excess of three. Cross-linking occurs in the usual way, copolymerized acid (e.g. methacrylic acid) acting as a catalyst for both etherification and cross-linking.

Typical resin recipes (before reaction with formaldehyde) might be:

(1) Styrene 82·5 per cent, acrylamide 15 per cent, methacrylic acid 2·5 per cent, by weight.
(2) Styrene 38·5 per cent, ethyl acrylate 44 per cent, acrylamide 15 per cent, methacrylic acid 2·5 per cent.
(3) Methyl methacrylate 25 per cent, ethyl acrylate 60 per cent, acrylamide 15 per cent.

Hardness decreases in the order (i) > (ii) > (iii). Flexibility increases with decreasing hardness.

There are other ways of producing cross-linked acrylic resins (see below and Chapters 14, 15 and 16). This route is described here because it makes use of nitrogen resin chemistry.

## Paints based on nitrogen resins

### U/F and M/F finishes

Since the cross-linking condensation reaction of U/F and M/F resins proceeds at temperatures of 90–180 °C and, with higher levels of stronger acid, at temperatures above 15 °C, two types of finish can exist:

(i)  the one-pack stoving finish
(ii)  the two-pack cold-curing finish

(i) STOVING FINISHES. These finishes have dominated the markets for metal coatings applied in the factory. With the resin properties already described – hardness and good colour – come the disadvantages of brittleness and poor adhesion. It is therefore necessary to plasticize the resins and this is usually done by blending with compatible alkyds, saturated polyesters or acrylic resins. 5–40 per cent nitrogen resin is used in the blend.

Short or medium oil alkyds, based on drying or non-drying oils, are used in a wide range of general purpose finishes. Specially formulated saturated polyesters are used where long term outdoor durability is required (e.g. pre-painted strip steel for cladding industrial buildings) or good chemical and corrosion resistance are needed (e.g. on washing machines). Maximum durability is obtained with polyesters made from *iso-* rather than *ortho-*phthalic acid and polyols containing no hydrogen on the carbon atom two places away from the hydroxyl group (the $\beta$ carbon atom), e.g. neo-pentyl glycol (see p. 167).

The acrylic resins used are flexible copolymers containing at least one 'hard' and one 'soft' monomer, a hydroxyl-containing monomer and possibly an acid monomer (see p. 141 for examples). They may be used as solutions in solvent or as non-aqueous dispersions. They produce extremely durable motor car finishes, which can be polished to some extent.

In all cases acidity, even in the resin, catalyses the cross-linking process and combination occurs by a condensation reaction between methylol groups in the nitrogen resins and hydroxyl groups in the plasticizing resin:

$$-CH_2OH + HO- \longrightarrow -CH_2O- + H_2O$$

The higher cost of M/F resins is justified where lack of discoloration is required at baking temperatures above 150 °C (e.g. in white finishes), where water and chemical resistance must be excellent (e.g. in washing-machine finishes) and where the best outdoor durability is required (e.g. in car enamels). These improvements, particularly in resistance and durability, are thought to be due to the absence of the water-sensitive carbonyl groups found in U/F resins and the increased cross-linking potential. The latter factor accounts for the shorter hardening times of M/F resins. A typical stoving schedule for a nitrogen resin finish might be 30 minutes at 127 °C.

The relatively low molecular weights of the nitrogen and plasticizing resins allow medium solids at spraying viscosities. Much higher solids are obtained with HMMM and special lower molecular weight plasticizing resins. The solvents used in these finishes consist usually of some alcohol (probably butanol) and aromatic hydrocarbons, such as xylene. Aliphatic mineral spirits may be present in some primers. Storage stability is generally good, since little or no reaction occurs in the can at room temperature unless the acid content of the paint is high, when a slow viscosity rise on storage can occur.

Because of the water-solubility of HMMM resins, they can be used with water-reducible alkyds, polyesters or acrylics and with acrylic copolymer latices to produce water-based counterparts of the above stoving finishes. Such coatings are not usually free of small proportions of water-miscible co-solvents, e.g. 2-butoxy ethanol, and ammonia or amines (see p. 109).

(ii) COLD-CURING FINISHES. U/F and M/F resins etherified with butyl or propyl alcohols and catalysed with stronger acids such as *p*-toluene sulphonic acid ($CH_3 \cdot C_6H_4 \cdot SO_3H$), will give good wood finishes with hardening times of a few hours at room temperature. Hardening can be accelerated by force-drying at temperatures which do not distort the wood, e.g. 60 °C. On more stable wood-based substrates, such as chipboard or hardboard, drying of clear or pigmented acid-catalysed finishes can be accomplished in a few minutes at over 100 °C or with infra-red heating. M/F resins give better resistance to water, solvents and chemicals than U/F resins do, but are slower curing at room temperature. The nitrogen resin is plasticized with non-drying or semi-drying alkyds and a solution of acid catalyst is supplied in a separate container. The finish and catalyst are mixed in a convenient ratio, e.g. 10/1 by volume, at the beginning of the day and the catalysed material remains usable at least for the remainder of the working day and usually for much longer.

These finishes compete chiefly with nitrocellulose lacquers and, although they are harder and more resistant, suffer by comparison for drying performance during the *early* stages of drying, when dust pick-up can be heavy. N/C lacquers can become dust free in 1–2 minutes, while unadulterated acid catalysed (A/C) wood finishes take 5–10 minutes. Since any compatible resin may be used in the A/C finish, it has been possible to produce 'combination finishes', containing amino resins, alkyds *and* nitrocellulose, with compromise drying times and film properties. Use of the higher molecular weight polymer necessitates a drop from the medium solids (35–45 per cent) of full A/C finishes, though the user is still better off than with N/C lacquers (15–30 per cent solids).

### Finishes based on acrylic nitrogen resins

Unlike *U/F* and *M/F resins*, acrylic nitrogen resins need not be blended with another component to obtain flexibility; this can be built into the acrylic copolymer by appropriate choice of monomers (see example on p. 177). Thus resins with ester linkages in their main polymer chains (e.g. alkyds, polyesters) can be avoided. A cross-linked polymer network can be built up which is resistant to hydrolysis by alkalis in detergents. Coatings can therefore be formulated which are suitable for domestic appliances, being additionally hard but flexible.

This family of resins is also widely used for the formulation of paints for pre-painted coiled metal strip, which may subsequently be fabricated into caravan exterior skins, etc. The maximum amount of methyl methacrylate is included to give good durability, but flexible monomer is needed to allow the painted sheet to be bent and formed.

Best results are obtained if the only acid present is part of the copolymer. Domestic appliance finishes cure in 20 minutes at 177 °C and coil coatings in 1 minute, with the metal reaching a peak temperature of 230–240 °C. Temperatures can be reduced, with some loss in properties, by the addition of a little acid catalyst, which is usually required when the film is hardened with small amounts of M/F. Other resins used to modify the film are epoxies, polyesters and alkyds.

Like other acrylic copolymers, these self-crosslinking acrylic nitrogen resins may be produced as latices or in water-soluble form and formulated into water-based paints.

### Toxicity

Concentrations of formaldehyde in air of 2 p.p.m. or more are intensely irritating to man and provide an unacceptable working environment. Additionally, as a result of tests on rats at much higher concentrations, some strongly disputed conclusions have been drawn that formaldehyde might be a human carcinogen. The ACGIH in the USA have therefore set a time-weighted average TLV of 1 p.p.m. and the H and SE in the UK have fixed a long term exposure limit of 2 p.p.m. Steps are being taken by all suppliers to minimize the free formaldehyde contents of all resins and coatings. Good extraction and working practices maintain workplace concentrations below the OEL or TLV. The nose and eyes are extremely good detectors of potentially hazardous concentrations of formaldehyde, as mentioned above.

Typical formulations based on nitrogen resins are given on pp. 181–82.

| Silver grey stoving enamel for motor cars | | | Clear acid-catalysed wood finish | |
|---|---|---|---|---|
| | wt % | | | |
| Aluminium paste (65% Al) | 2·18 | | | |
| Carbon black | 0·07 | **Pigments** | | |
| Synthetic yellow iron oxide | 0·04 | | (a) *Clear coating* | wt % |
| Synedol 2263 XB | 69·12 | **Resin Solutions** | Beetle alkyd resin BA509 | 48·0 |
| Synresine ME 2070 | 16·42 | | Beetle urea resin BE678 | 24·5 |
| 'Cellosolve' acetate | 2·30 | | | |
| Toluene | 6·88 | **Solvents** | Butanol | 27·5 |
| n-Butanol | 2·30 | | | ——— |
| | | | | 100·0 |
| | | | (b) *Activator* | |
| 1% silicone oil in xylene | 0·69 | **Additive** | p-toluene sulphonic acid | 20·0 |
| | | | Distilled water | 2·0 |
| | | | Isopropanol | 78·0 |
| | ——— | | | ——— |
| | 100·00 | | | 100·0 |
| | ——— | | | ——— |

Thin with xylene: n-butanol: 'Cellosolve'
acetate, 65 : 30 : 5 for spray application
at approx. 30% solids. Stove 30 mins
at 127 °C

Activate 100 parts coating:
2 parts activator.
Spraying solids: 40%. Air-dry,
or force dry 10 min at 80 °C.

*Synedol 2263 XB* is an acrylic copolymer (containing hydroxyl groups) at 50% in
xylene and n-butanol. *Synresine ME 2070* is an *iso*-butylated M/F resin at 70% solids
in *iso*-butanol. Both resins are made by DSM Resins BV.
*Beetle BA 509* is a short-medium semi-drying oil alkyd at 50% in xylol/butanol. *Beetle
BE 678* is an *iso*-butylated U/F resin at 66% in *iso*-butanol. Both resins are made by
British Industrial Plastics Ltd.

Water-reducible M/F: alkyd stoving finish

|  | wt % |  |
|---|---|---|
| Rutile titanium dioxide | 25·84 | **Pigment** |
| Amoco WS-3823 alkyd | 29·04 ⎱ | **Resins** |
| Cymel 301 | 3·94 ⎰ |  |
| Dimethylethanolamine | 1·12 | **Neutralizing base** |
| Water | 40·06 | **Solvent** |
|  | ───── |  |
|  | 100·00 |  |
|  | ───── |  |

Thin with water to approx. 30% spraying solids.
Stove 30 min at 175 °C.

*Amoco WS-3823* is a short oil tall oil/isophthalic acid/neopentyl glycol/trimellitic anhydride alkyd of acid value 36, which is at 80% solids in 2-butoxy ethanol. The formulation is available from Amoco Chemicals. *Cymel 301* is an HMMM resin supplied at 100% solids by American Cyanamid.

# Fourteen

# Epoxy coatings

Under this heading come a wide variety of finishes based on the epoxy resins. These resins involve in their preparation, and later in their cross-linking, the reactions of the epoxide ring, which are described in Chapter 4. A brief revision of these reactions would be worthwhile before proceeding further.

The variety of finishes is so wide, that their uses will be described later in detail, while the characteristic 'epoxy' properties will be more easily understood when the resin structures are seen. Let us proceed to the resins.

## Epoxy resins

These are usually prepared from epichlorhydrin and a dihydroxy compound (see p. 184). The latter is usually a diphenol – in particular the compound called bisphenol A – though it can be a dihydric alcohol. Two reactions of the phenolic hydroxyl bring about the polymerization:

(1) Condensation with chlorine to eliminate HCl.
(2) Addition to epoxide, opening the ring. It should be noted that this produces one hydroxyl group.

Thus this type of epoxy resin contains a maximum of two epoxide rings *at the ends* of the molecule and a number (which may be zero) of hydroxyl groups along the chain. These groups make the resin a polar one and ensure good adhesion to polar or metallic surfaces. The next thing to note is that the polymer chain contains only carbon-carbon and ether linkages. Both are very stable. We have seen the C—C stability in acrylic polymers, and in the preparation of nitrocellulose we noted that very severe conditions were required to break the ether linkage. Both linkages are much stronger than ester linkages, which are saponified by alkali. Epoxy resins, therefore, have good chemical resistance. Again, the phenolic hydroxyls, which lead to poor colour in phenolic resins, are entirely converted to ether links in epoxy resins, which are therefore of good colour. Epoxy resins on their own are low molecular weight, relatively brittle materials and are useful film-formers only when cross-linked with other molecules. Cross-linking takes place

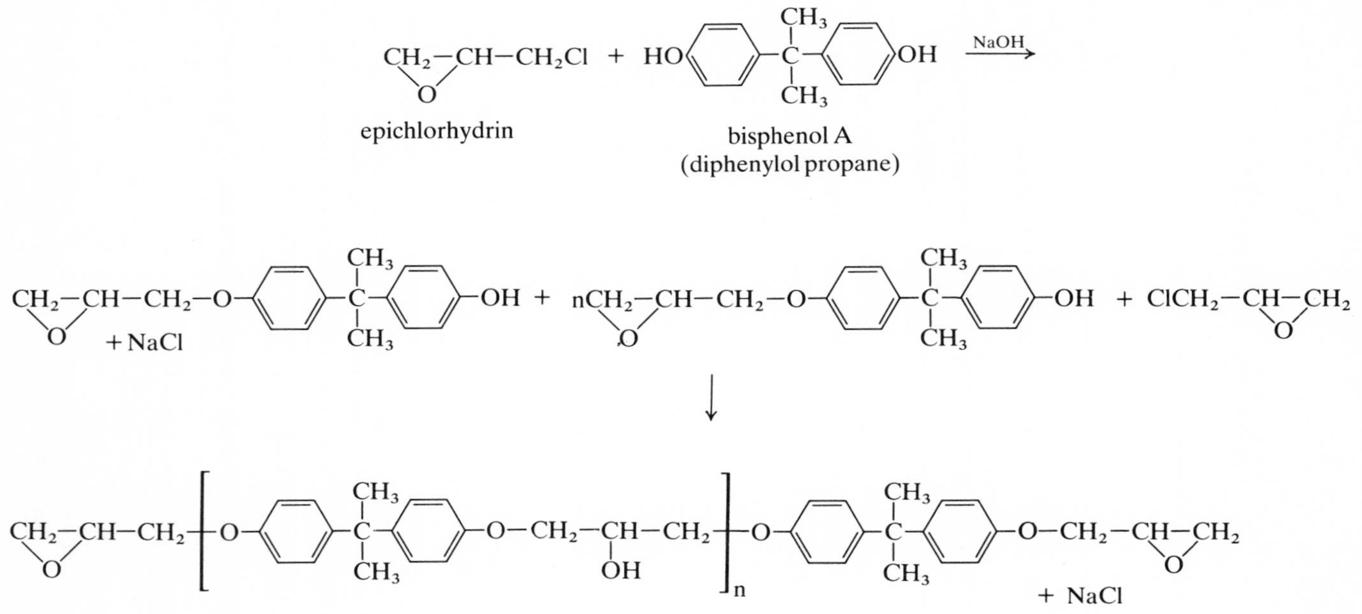

$$CH_2\!-\!CH\!-\!CH_2Cl \;+\; HO\!-\!\!\underset{\text{bisphenol A}}{\phantom{}}\!\!-OH \xrightarrow{\;NaOH\;}$$

epichlorhydrin      bisphenol A
(diphenylol propane)

$$CH_2\!-\!CH\!-\!CH_2\!-\!O\!-\!\cdots\!-OH \;+\; nCH_2\!-\!CH\!-\!CH_2 \;+\; \cdots\!-OH \;+\; ClCH_2\!-\!CH\!-\!CH_2 \longrightarrow$$

$$+ NaCl$$

$$\left[\; \cdots \;\right]_n \;+\; NaCl$$

through the reactive epoxide rings and hydroxyl groups. These are well separated by 3 carbon atoms, 2 oxygen atoms and 2 benzene rings. Because the cross-links *cannot* be closely spaced the resins give flexible cross-linked films. However, the aromatic epoxy resins strongly absorb u.v. light and are degraded by it, leading to chalking if used in topcoats outdoors.

Unmodified epoxy resins vary from the viscous liquid 'diepoxide 0' (molecular weight 340 and with $n = 0$ in the general formula above), to solid polymers of molecular weights up to 8000 ($n = 26$). The resins become solid at molecular weights above 700. Like all polymers, epoxy resins contain a mixture of molecular species. Not all molecules are terminated at both ends by epoxide groups. Low molecular weight diluents, containing perhaps one epoxy ring, are often included to make the 'diepoxide 0' types less viscous. Resins containing these diluents are described as 'modified' epoxies. The solvents for epoxy resins are given in Tables 2 and 4, Chapter 9. It should be noted that the petroleum hydrocarbons are non-solvents.

Three types of alternative epoxy resin are worth a special mention. First is the *epoxy novolac* resin, which contains more than two epoxide rings. Such a resin is made from a phenolic novolac (see p. 164) and epichlorhydrin and usually contains three to four epoxy rings per molecule:

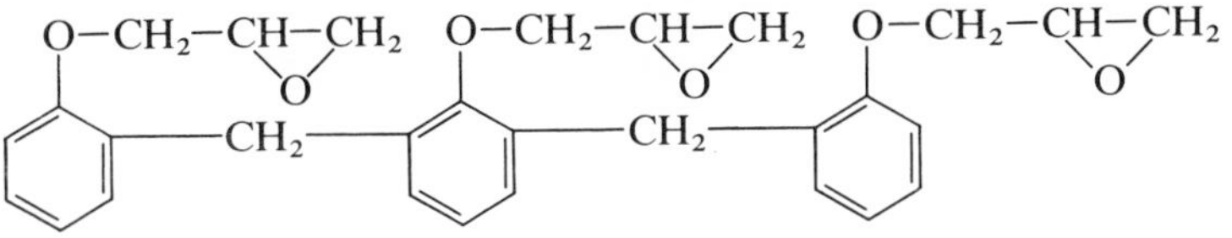

$$O-CH_2-CH-CH_2 \quad O-CH_2-CH-CH_2 \quad O-CH_2-CH-CH_2$$

Second is the *cyclo-aliphatic epoxy* resin, in which the epoxide ring is incorporated onto one of the sides of a *cyclo*-hexane ring, e.g.

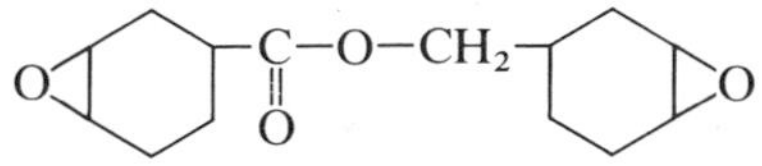

Finally, it is possible to produce an *acrylic epoxide resin* by copolymerizing glycidyl methacrylate

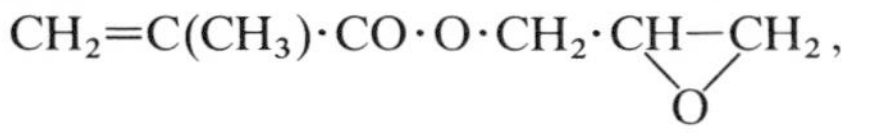

$$CH_2{=}C(CH_3)\cdot CO\cdot O\cdot CH_2\cdot CH-CH_2,$$

with other acrylic monomers.

We can regard epoxy resins in two ways: as cross-linking resins and as polyhydric alcohols. Let us consider the second aspect first, since it leads to the possibility of further resins.

**Epoxy resins as polyols**

(i) EPOXY-ESTER RESINS. These can be considered as alkyd equivalents. They are usually made by esterifying the epoxy and hydroxyl groups with oil fatty acids. The epoxy groups are more reactive to carboxyl than are hydroxyl groups and, in the process of reaction, create a hydroxyl group on the carbon atom adjacent to the ester linkage. Thus each ideal bisphenol A epoxy polymer molecule has two sites of epoxide reactivity and *n plus two* sites of hydroxyl reactivity.

If *n* is too high, the partially esterified epoxy molecule will have a high functionality of oil side chains. Should the oil be prone to heat bodying (e.g. a drying oil), gelation could easily occur as the temperature is raised to complete the esterification. The oil side chains would cross-link the epoxy backbones. For this reason, *n* is usually kept below 4 (molecular weight 1400), though non-drying or semi-drying oils can be used with resins up to *n* = 9 (molecular weight 2900).

Acid values of 0–10 are usual. Oil length is described as the fraction of potential ester linkages formed. For example, a resin with average value of *n* = 3·7 can form an average of 7·7 linkages (4 from the epoxy groups) per molecule. If only three linkages are made with fatty acids, this is a fraction of 0·4 of the possible total, or 40 per cent. 30–50 per cent esterification is described as short oil, 50–70 per cent as medium oil and 70–90 per cent as long oil length. Properties vary with oil length very much as they do in alkyds. Resins of a given oil length dissolve in solvents suitable for alkyds of similar oil length. Uses are also similar, but epoxy-ester resins have better adhesion and chemical resistance than alkyds and the stoving versions are harder for the same flexibility.

The drying oil fatty acid chains can be partially maleinized to permit emulsification when the carboxyls have been partially neutralized and the resin dispersed in water/solvent mixtures. These waterborne epoxy-esters can be used as binders for anodic electrodeposition primers (see p. 110).

(ii) EPOXY-ALKYDS. Epoxy resins may be used as polyols in alkyd preparation. Since the lowest functionality possible – even with liquid epoxies – is four (two epoxide groups), they are usually used with other polyols such as glycerol and the functionality is previously reduced to three by reaction with fatty acids. Epoxy resins of molecular weight up to 1400 have been used in alkyds. The usual epoxy properties are conferred on the resin, the extent of change being dependent on the amount of epoxy resin incorporated.

## Epoxy coatings

These may contain the epoxy resins, epoxy-esters, epoxy-alkyds or other possible epoxidized resins. Both stoving and air-drying variants are possible.

### Stoving and u.v. curing finishes

Short-medium oil epoxy-esters can be used in stoving finishes, with or without nitrogen resins, in place of the corresponding alkyds. They make excellent primers and surfacers, especially for use under motor-car finishes.

The epoxy resin molecules cross-link with each other and with a variety of other resins.

(i) CROSS-LINKING WITH RESINS CONTAINING HYDROXYL GROUPS. In this category come the phenolic and amino resins. The principal reaction is addition of hydroxyl to the epoxide ring (as in the preparation of epoxy resins), but condensation reactions between the hydroxyl groups occur to a lesser extent.

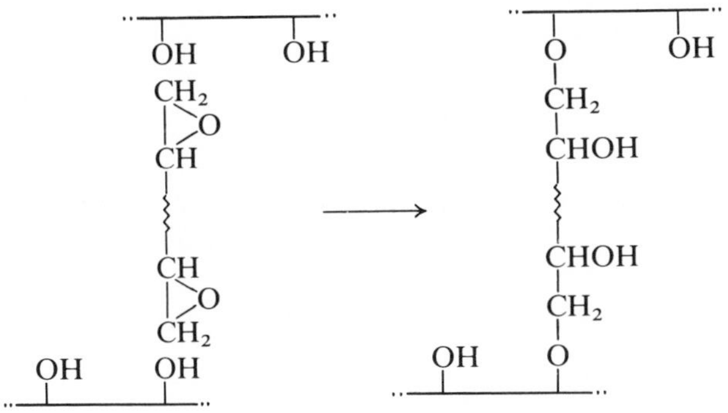

(only reacting groups shown)

*Epoxy-phenolic finishes* are among the most chemically resistant known. 20–30 per cent phenolic resin is used with epoxy resin of molecular weight 1400 and the stoving schedule is 20 min at 180–205 °C. Phenolic resins alone give excellent chemical resistance, but cross-linking with epoxy resin improves adhesion and impact resistance. The poor colour of the coatings is due to the phenolic resin. The two resins may be partially reacted together before being used in a paint. Since part of the hardening reaction will then have been carried out before painting, shorter stoving times are possible for the paint user. Also, resins that are incompatible on cold-blending can be made compatible by partial combination in the resin kettle. This broadens and cheapens the range of phenolic resins that can be used. The main uses for these finishes are as can and drum linings.

*Epoxy-U/F resin* combinations are almost as chemically resistant as epoxy-phenolics and have better colour. They make excellent corrosion-resistant primers for coating of coiled metal strip. $5 \mu$m films are stoved for 30–60 seconds to a peak metal temperature of 240–250 °C.

(ii) CROSS-LINKING WITH RESINS CONTAINING CARBOXYL GROUPS. Esterification of the epoxy groups is the means of cross-linking. This method provides another means of obtaining a *thermosetting acrylic finish*.

$$CO \cdot OH \quad CO \cdot OH$$
$$CH_2$$
$$\overset{\diagdown}{\underset{\diagup}{\phantom{.}}} O$$
$$CH$$
$$\vdots$$
$$CH$$
$$\overset{\diagdown}{\underset{\diagup}{\phantom{.}}} O$$
$$CH_2$$
$$CO \cdot OH \quad CO \cdot OH$$

$$\longrightarrow$$

$$CO \qquad CO \cdot OH$$
$$O$$
$$CH_2$$
$$CHOH$$
$$\vdots$$
$$CHOH$$
$$CH_2$$
$$O$$
$$CO \cdot OH \quad CO$$

The acrylic copolymer is made to contain about 4–12 per cent of acrylic or methacrylic acid and the minimum number of carboxyl groups per molecule for cross-linking (3) is usually considerably exceeded. Stoving temperatures are 170–180 °C, or 150 °C if basic catalyst is added. Such combinations have limited shelf life, because of slow reaction in the can. Domestic appliance acrylics are usually as described in Chapter 13, often with minor inclusions of epoxy resin to upgrade resistance properties.

(iii) EPOXY RESIN/M/F RESIN/ALKYD FINISHES can be considered as reacting by both routes, since the unreacted acid in the alkyd may take part. Epoxy resin of molecular weight 900 ($n = 2$) is used, since alkyd/epoxy resin compatibility is poor at higher molecular weights. Stoving is at temperatures of 150 °C and above. Improvements over alkyd/melamine finishes occur in adhesion, flexibility, chemical and water resistance and mar and abrasion resistance. Uses are similar, particularly where finishing industrial and domestic equipment is concerned, since the excellent adhesion of the epoxy resin can allow the user to omit a primer.

(iv) POWDER COATINGS. Powder coatings are not necessarily based on epoxy resins; there are polyester, polyurethane, acrylic and other types. However, epoxy powders have dominated powder coating since serious growth began in 1964 and still have such a large share of sales that it is appropriate to discuss the topic in this chapter.

A powder coating is made by mixing molten resin, pigments, cross-linking agent and additives uniformly, extruding the melt, breaking up the solidified extrusion and finally milling to produce a powder with a particle size distribution of approximately 10–100 $\mu$m. The pigmented powder is then applied, either by immersing the article in a fluidized bed of powder (powder kept mobile by the passage of air through it) or, more usually, by special electrostatic spray guns. The film is formed when the particles are sintered in an oven and cross-linking also occurs.

Powder coatings do not readily give satisfactory films at thicknesses below about 45 μm, and require high temperatures for stoving and special equipment for handling and application. However, they form good-looking, tough thick coatings and contain no solvent. Wastage is very low, but colour changes in one application unit can lead to speckled finishes, unless scrupulous changeover procedures are followed.

To be suitable for powder coatings, resins should melt for pigmentation at temperatures well below that required for cross-linking, should be brittle and amenable to dry milling when cold, should be stable in the presence of cross-linker when stored at room temperature and, at stoving temperatures, should melt and fall rapidly in viscosity before cross-linking. Epoxy resins admirably meet these requirements.

Cross-linkers for epoxy powders can be nitrogen resins or polyamides (see p. 192), but two more popular types require special mention.

Dicyandiamide (see p. 175) and its derivatives cross-link at 180 °C by reaction of the amino and imino (=NH) groups with the epoxy ring:

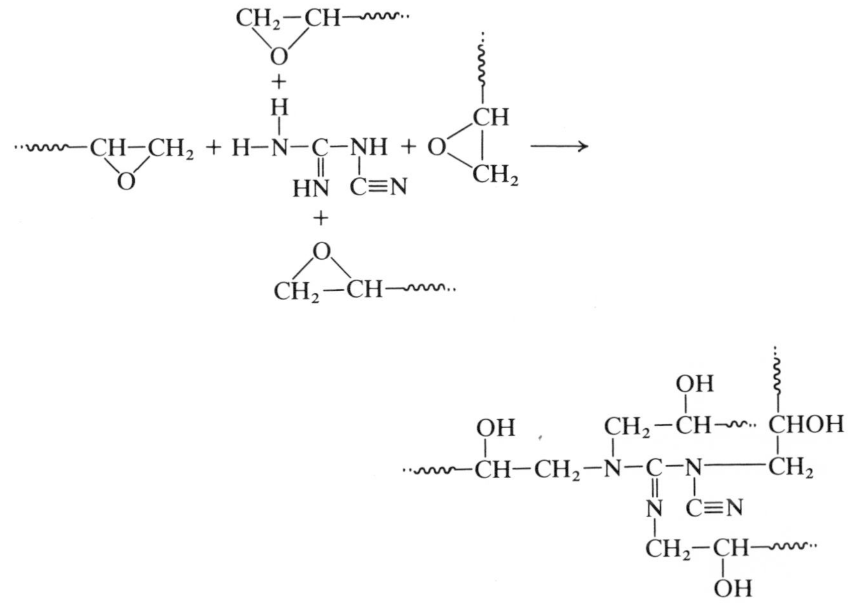

Anhydrides cross-link epoxy resins by reacting first with the hydroxyl groups. Carboxyl groups are produced in the process and these then react with the epoxy ring. Stoving is at 180–200 °C.

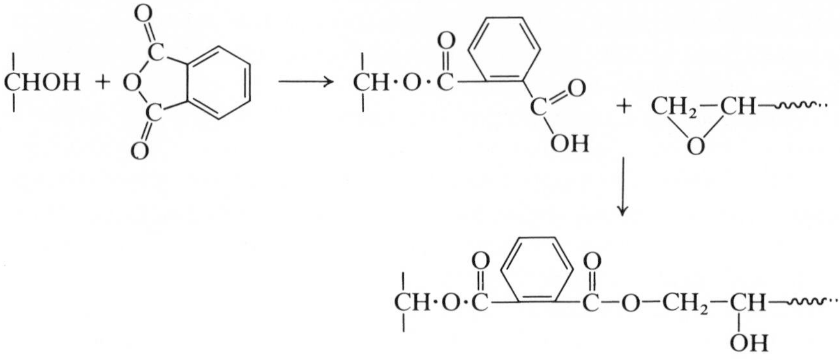

Both types of cross-linking reaction can be promoted by the addition of catalysts.

(v) CATIONIC SELF-POLYMERIZATION AND U.V. CURING. The epoxy ring

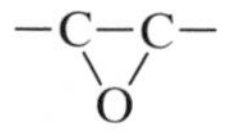

looks somewhat like a carbon–carbon double bond —C=C—, so it may not be surprising to learn that it is capable of addition polymerization by a chain reaction. Free radicals are not, however, thought to be involved. Normal free radical initiators (e.g. organic peroxides, Chapter 16) do not bring about the polymerization. Polymerization is brought about by Lewis acids (e.g. $BF_3$), but it is so rapid and vigorous that they cannot be used as paint activators. Instead, we require complex salts with large unstable organic cations and large generally inorganic anions, e.g.

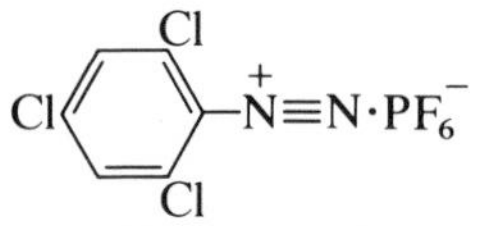

2,4,6-trichlorobenzene diazonium          di (*p-t*-butylphenyl) iodonium
hexafluorophosphate                       hexafluoroantimonate

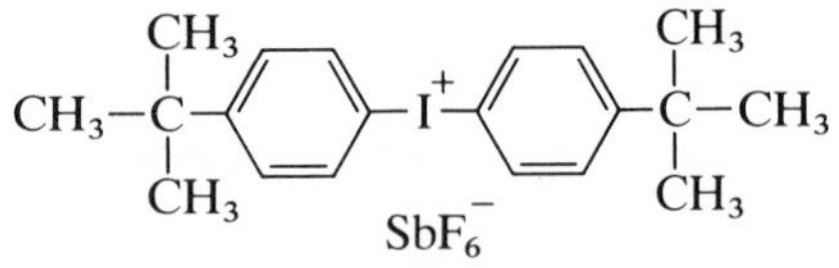

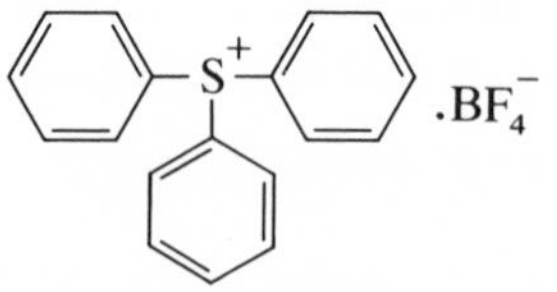

triphenyl sulphonium tetrafluoroborate

All of the above 'onium' salts are relatively stable when included in epoxy coatings but, on exposure to ultra-violet light, they are decomposed to form reactive cations. A full understanding of cationic polymerization processes has yet to be obtained. We do know that bases, water and smaller anions, e.g. Cl⁻, that readily attack positive sites in organic compounds, inhibit polymerization. We may illustrate the cationic polymerization of epoxies by assuming that initiation occurs via a proton

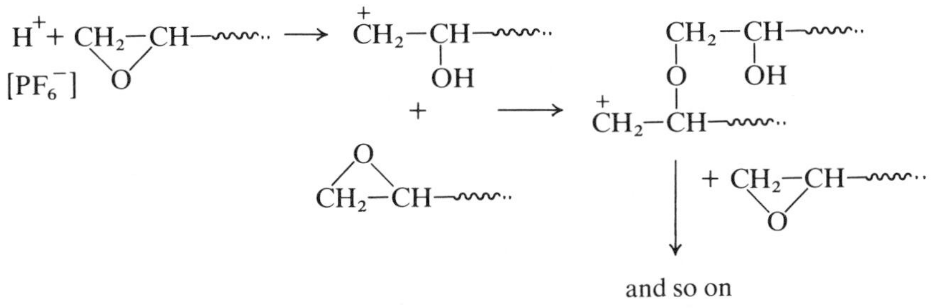

and so on

After decomposition of the photo-initiators by u.v. light, polymerization is very rapid and films may be substantially cured in less than a second. Further polymerization can occur at room or higher temperatures after u.v. illumination has ceased.

All the types of epoxy resins mentioned on p. 185 may be used to form coatings with good adhesion to metal and are particularly suitable for the exterior surfaces of cans. *Cyclo*-aliphatic epoxides give fastest cure, novolacs have the highest epoxy functionality and dihydric alcohol diepoxides impart flexibility. Viscosity is reduced by monoepoxides and small proportions of alcohols (which terminate the chain growth). The polymerization is not inhibited by oxygen (see Chapter 16) and the coatings need not contain any significant amount of volatile material, i.e. they are almost 100 per cent solids.

**Finishes curing at normal temperatures**

Long oil epoxy-esters, based on drying oil fatty acids, can be used in varnishes, but with poor resistance to chalking out-of-doors, they do not compete with long oil alkyds in decorative paints.

Epoxy resins proper may be cross-linked by reaction with chemicals containing amino groups. The finishes produced may be classified as follows:

(i) FINISHES CURED BY POLYAMINES. The addition reaction, between epoxide ring and primary and secondary amino groups that are not directly attached to a benzene ring, occurs at temperatures above 15 °C and provides a means of curing the resins at room temperature. To cross-link diepoxide resins, the amine must have a functionality of at least three; that is to say, at least three amino hydrogen atoms.

Simple polyamines such as ethylene diamine, $H_2N \cdot CH_2CH_2 \cdot NH_2$ (functionality 4) and diethylene triamine, $H_2N \cdot CH_2 \cdot CH_2 \cdot NH \cdot CH_2 \cdot CH_2 \cdot NH_2$ (functionality 5) are used, being supplied as an 'activator' solution. This is added to the paint (which contains the epoxy resin) just before it is used. The amines are not catalysts, since they are consumed in the chemical reaction. Epoxy resins of low molecular weight (approx. 900) are preferred. Cross-linking might occur thus:

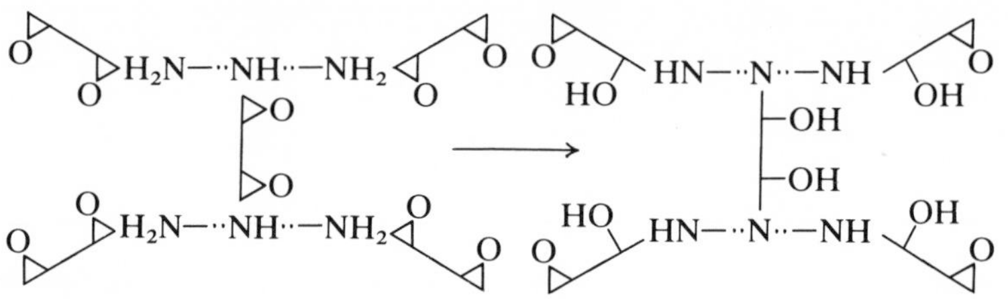

Such paints are prone to surface exudation. The risk of getting this fault can be minimized by allowing the paint to stand 12–16 hours before it is used. This is quite safe, since pot-lives are arranged to be about 48 hours at normal temperatures. An alternative, more reliable method is to pre-react some epoxy resin with an excess of amine, so that an adduct containing unreacted amino hydrogen atoms is produced.

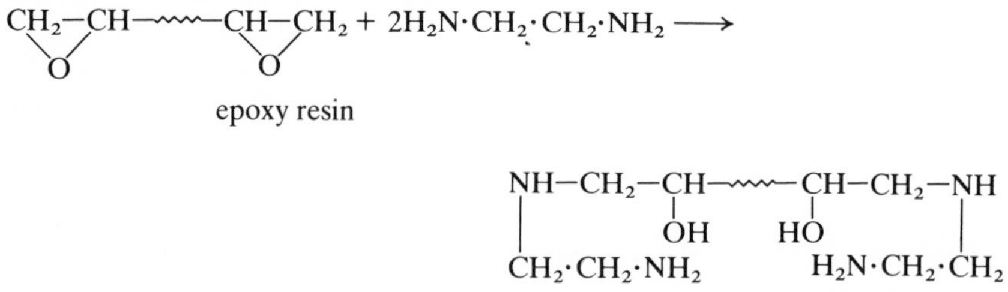

epoxy resin

adduct (functionality 6)

Ester and ketone solvents, which react with amines, must not be present. The adduct is a polyamine and can therefore be used to cross-link epoxy resins. It has advantages in that it is odourless (simple amines are unpleasant to handle) part-reacted (which speeds up drying) and free from simple amines (which cause exudation). It can be seen now why it is an advantage to age paints cured by simple amines: amine adducts are formed during the ageing.

(ii) FINISHES CURED BY POLYAMIDE RESINS. Polyamide resins are prepared from polyamines and dimer fatty acids. Dimer fatty acids are produced from unsaturated drying (or semi-drying) oil fatty acids by Diels–Alder reaction.

A possible dimer acid might be:

$$CH_3 \cdot (CH_2)_5 - CH - CH - CH = CH - (CH_2)_7 - COOH$$
$$CH_3 \cdot (CH_2)_5 - CH \quad\quad CH - (CH_2)_7 - COOH$$
$$CH = CH$$

from dehydrated castor oil fatty acids

This can be thought of as HOOC⎯⌁⎯COOH and will react with polyamine to form polyamide resin:

$$nHOOC⎯⌁⎯COOH + nH_2N⎯\cdots⎯NH_2 \longrightarrow$$

$$HOOC⎯⌁⎯CO \cdot NH⎯\cdots⎯NH \cdot CO⎯⌁⎯CO \cdot NH⎯\cdots \text{etc.}$$

$$+ 2n - 1H_2O \quad\quad \underbrace{\quad\quad\quad\quad}_{\substack{\text{amide} \\ \text{linkage}}}$$

If *excess* amine is used and some triamine is included, branched polyamides can be produced, which contain unreacted amino groups at the ends of chains. They can be thought of as high molecular weight *polyamines* and as such are clearly suitable for curing epoxy resins.

The polyamide resins increase the flexibility of the film, the amine groups being well spaced in the large molecules, but, because amide groups have now become linkages in the polymer structure, alkali resistance is slightly reduced. Amides are hydrolysed by alkali, though not readily. Cure is slower and pot lives are longer. At air temperatures above 15 °C, the films are surface-dry in a few hours, but full hardness and resistance is only obtained in seven days. Initial force-drying (at 40–60 °C) will reduce this period. Pigment dispersion can be carried out in the polyamide resin solution: this may give easier dispersion in some cases.

Epoxy/polyamine or polyamide finishes are suitable wherever chemical resistance is required, chalking cannot occur (indoors) or can be accepted and stoving is not possible. The chemical resistance obtained does not equal that of epoxy-phenolic and epoxy-urea finishes, but it is good for a paint drying at atmospheric temperatures. Typical outlets are the maintenance painting of large industrial installations and equipment and the priming of aircraft. With a combination of hardness, flexibility and adhesion, the coatings are also suitable for wood and concrete (e.g. floor coatings). The inclusion of coal tar in the amine pack (amount approximately equal to the weight of epoxy resin) upgrades water and corrosion resistance, e.g. for undersea protection of metal (colour ranges are limited).

(iii) SOLVENTLESS FINISHES. It has been pointed out that the simplest epoxy 'resins' are liquids. Undiluted 'diepoxide O' has a viscosity of 36–64 poises at 23 °C, but this viscosity can be lowered further by reactive diluents. The 'modified' resin can have a viscosity as low as 5 poises. The diluents have an epoxide functionality of only one, so they do not contribute to cross-linking, but they are chemically bound into the final cross-linked film. Since these liquid epoxies may be cured by liquid polyamines or polyamides, no further solvent is required for application. Thus the entire liquid paint can be converted to cross-linked solid without heating, so that it may be called 'solventless', or a '100 per cent solids' finish.

The main outlet for such a finish is an anti-corrosive lining for storage tanks, both on land and in oil tankers. Maximum corrosion resistance is obtained only (*a*) if the steel is cleaned by grit blasting and (*b*) if it is covered by about 125 $\mu$m of coating. The grit-blasting process leaves a pitted 'profile' on the steel. Ideally, the maximum amplitude of this profile should be determined and a measured thickness of coating equal to three times that amplitude should be applied. In practice, it is usually necessary to apply 200–250 $\mu$m of coating to protect the high spots adequately. Such thicknesses are only achieved in four or more coats with finishes containing solvent and each coat must be allowed to dry partially before the next is applied. Labour costs are naturally high and the process is long. An oil tanker in dock would at the same time be running up high docking charges. This market is therefore made for a 100 per cent solids epoxy finish, which can be applied at 100–400 $\mu$m in one coat, even though the paint itself is very expensive (since solvents are relatively cheap and resins – especially epoxies – are not). Such thick coatings are prevented from sagging by inclusion of mineral thickeners. The viscous finish is applied by airless spraying, where the coating is forced through a nozzle under high pressure. Large areas are coated in a very short time.

In many solventless systems (see 'polyesters', Chapter 16), there is an appreciable volume contraction in the change from liquid linear polymer and co-reactants to solid cross-linked polymer. This can put the adhesive bond to the substrate under great strain and can cause loss of adhesion, pulling away from edges and related defects. Solventless epoxy systems are remarkable in that they show negligible contraction.

The solventless epoxy film shows typical epoxy finish properties, but is inevitably less flexible than usual because (*a*) the films are thicker and (*b*) the close spacing of the reacting groups leads to a high density of cross-links. The film is, however, very tough and does not shatter easily.

Because the reactants are not diluted in solventless coatings, pot-lives are short (0·5–2 hours). Because of short pot-lives, special twin-feed airless spray equipment is often used. Aromatic polyamines are preferred for curing, so that hardening times of 4–12 hours can be obtained even at 0–10 °C. This better low temperature cure cannot be obtained in outdoor topcoats, where colour is important, since the aromatic polyamines discolour.

As well as the ingredients already mentioned, tertiary amino phenolic catalysts, such as tri (dimethylaminomethyl) phenol,

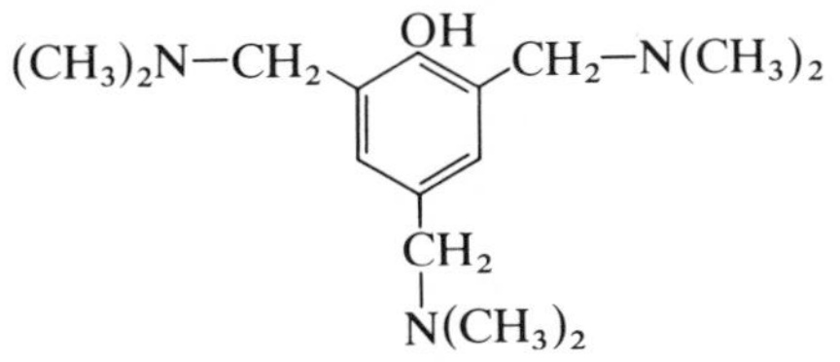

$$(CH_3)_2N-CH_2 \quad \overset{OH}{\underset{CH_2}{\bigcirc}} \quad CH_2-N(CH_3)_2$$
$$\underset{N(CH_3)_2}{|}$$

are often included. These produce polymerization of the epoxy resin with itself:

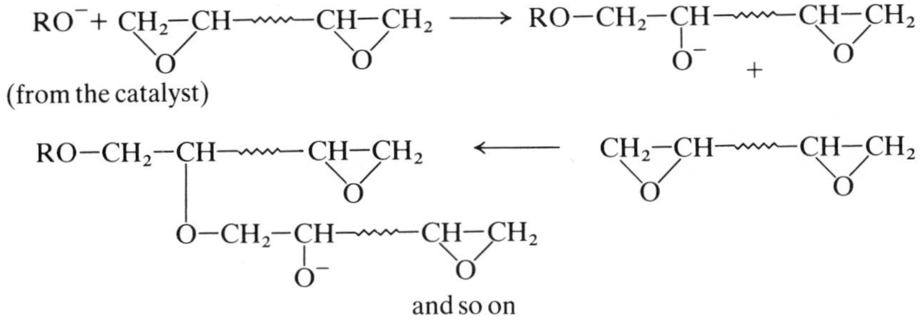

$$RO^- + CH_2-CH\text{---}CH-CH_2 \longrightarrow RO-CH_2-CH\text{---}CH-CH_2$$

(from the catalyst)

$$RO-CH_2-CH\text{---}CH-CH_2 \longleftarrow CH_2-CH\text{---}CH-CH_2$$
$$O-CH_2-CH\text{---}CH-CH_2$$

and so on

About seven epoxy groups are reacted in this way for each nitrogen atom in the catalyst. Phenols are also included and act as accelerators in the curing reaction. Silicone resins may be added as flow agents and dibutyl phthalate has a plasticizing action.

## Water-based epoxy coatings

Although the high solids, solventless and powder versions of epoxy coatings have created wider interest, it is possible with epoxies, as with most resin systems, to produce water-based versions. Usually the epoxy resin is emulsified with suitable surfactants. The co-reactive material can be, for example, a U/F resin emulsified with it to produce a stoving composition, or a water-soluble polyamine supplied in a second pack for room temperature cure.

Some examples of different types of epoxy paints are given overleaf:

---

Epoxy-phenolic clear interior can coating

|  | wt % | |
|---|---|---|
| Epikote 1007 | 24·0 ⎫ | **Resin** |
| Bakelite 100 | 8·0 ⎭ | |
| 1-Methoxy propan-2-ol | 45·3 ⎫ | **Solvent** |
| Shellsol A | 22·6 ⎭ | |
| Phosphoric acid | 0·1 | **Catalyst** |
|  | 100·0 | |

---

Apply by roller coater at 32% solids.
Stove 15 min at 200 °C.
*Bakelite 100* is a phenolic resole from Bakelite GmbH.
The following are products of Shell Chemicals:
*Epikote 1007* is a bisphenol A epoxy resin of molecular weight 2900.
*Shellsol A* is an aromatic hydrocarbon blend, BR 162–180 °C.

---

Anodic electrodeposition primer

|  | wt % | |
|---|---|---|
| Kronos RN 57 titanium dioxide | 2·68 ⎫ | **Coloured pigment** |
| Sterling carbon black | 0·02 ⎭ | |
| Strontium chromate | 0·39 ⎫ | **Anti-corrosive** |
| Oncor F-31 | 0·39 ⎭ | **pigment** |
| Hydrite 10 clay | 0·38 ⎫ | **Mineral thickeners** |
| Ben-a-gel | 0·04 ⎭ | |
| Epikote resin ester DX-38 | 11·19 | **Resin** |
| Shellsol A | 3·34 ⎫ | **Organic solvent** |
| 1-Methoxy propan-2-ol | 0·83 ⎭ | |
| Triethylamine | 0·80 | **Neutralizing base** |
| Demineralized water | 79·94 | |
|  | 100·00 | |

---

Apply 25 μm by electrodeposition on anode from above paint at 15% solids.
Stove 30 min at 180 °C.
*Oncor F-31* is basic lead silico chromate from NL Industries.
*Epikote resin ester DX-38* is the reaction product of Epikote DX-20 epoxy resin from Shell Chemicals, in which the epoxy rings are reacted with a styrene/methacrylic acid copolymer, further reaction is carried out with linseed oil fatty acids and finally the product is reacted with maleic anhydride to an acid value of 80 mg KOH/g.

TWO-PACK SOLVENTLESS FINISH

| Finish | | | Activator | |
|---|---|---|---|---|
| | wt % | | | wt % |
| Rutile titanium dioxide | 6·4 | **Pigment** | Hardener HY830 | 48·5 |
| Barytes | 39·8 | **Extender** | Hardener HY850 | 51·5 |
| Bentone 27 | 3·2 | **Thickener** | | ——— |
| Araldite GY 250 | 47·2 | **Resin** | | 100·0 |
| Dibutyl phthalate | 3·4 | **Plasticizer** | | ——— |
| | ——— | | mixing ratio: | |
| | 100·0 | | Finish/activator, 3·3/1 by wt. | |
| | ——— | | | |

Drying times: tack free 4–5 h, hard 8–14 h at 7°–20 °C.
Pot life: 1 h.
*Araldite GY 250* is an unmodified liquid epoxy resin, viscosity 225–275 poises at 21 °C, made by CIBA (A.R.L.) Ltd.
*Bentone 27* is a treated clay from F. W. Berk and Co. Ltd.
*Hardeners HY830 and HY850* are aromatic amines of 90 and 350 poises respectively at 21 °C. *HY850* also includes an accelerator. Both are made by CIBA (A.R.L.) Ltd.

# Fifteen

# Polyurethanes

Polyester finishes contain polyester resins, which, in turn, contain predominantly ester linkages. Polyurethane finishes need not contain polyurethane resins and the urethane linkage is not necessarily predominant in the dry paint film. However, all polyurethane finishes contain isocyanates or their reaction products, so the reader should be familiar with isocyanate reactions before proceeding further with this chapter.

Since there is a variety of isocyanate reactions, the formulator has considerable scope. So far, five different approaches have been used: three have led to one-pack paints and two to two-pack paints.

## One-pack paints

### (a) Urethane oil and alkyd finishes

Oils can contain hydroxyl groups in the fatty acid portion (e.g. castor oil), or can be converted to diglycerides and monoglycerides when they are heated in the presence of *poly*hydric alcoh*ols* (polyols). By addition of diisocyanate, the hydroxylic oil can be chemically 'bodied', the molecules being linked together by urethane linkages, e.g.:

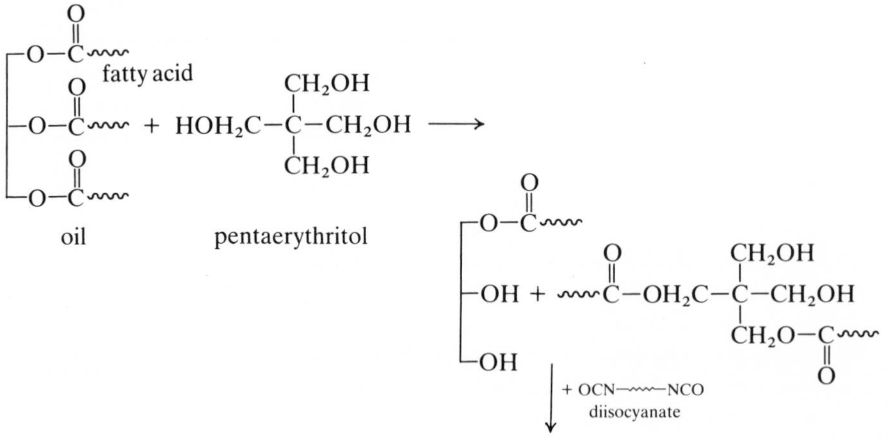

Urethane oil

An excess of isocyanate groups is not permitted, so that the final product, a *urethane oil*, contains no unreacted —NCO. It is therefore stable to moisture, non-toxic and differs from bodied oil in that it contains urethane linkages and can be much higher in molecular weight. The amount of diisocyanate used and the molecular weight of the product, vary with the extent of conversion of oil to mono- and diglyceride and with the relative proportions of these two species. The product has an oil length which is either 'long' or 'medium' (see the definition of the oil length of an alkyd).

A *urethane alkyd* is an alkyd in which some of the dibasic acid is replaced by diisocyanate. The ester links are formed first in the usual way, the diisocyanate is added and the remaining hydroxyls are reacted at 80–95 °C to form the urethane linkages (see p. 200).

As with the urethane oils, no unreacted isocyanate groups are permitted in the final alkyd.

Ordinary drying-oil alkyds are restricted in molecular weight by the difficulty of getting esterification to occur when the concentration of carboxyl and hydroxyl groups becomes low near the end of the preparation. Esterification can be continued only by pushing up the temperature. This now rises into the region where heat-bodying of the unsaturated fatty acid chains can occur. The coupling of fatty acid chains cross-links the alkyd polyester backbones and can lead to rapid irreversible gelation. Therefore the reaction must be stopped. Since urethane alkyds can be completed at low temperatures where heat-bodying cannot occur, high molecular weight molecules can be obtained. They dry faster than ordinary alkyds because (i) the molecules are initially bigger and (ii) they have a higher fatty acid functionality for oxidative drying.

Urethane alkyds are basically similar to ordinary alkyds and may be drying or non-drying, long oil or medium oil. Short oil urethane alkyds are theoretically possible, but are difficult to make. Irreversible gelation tends to occur in the reaction vessel.

Some authors apply the terms 'urethane alkyd', 'urethane oil' and 'uralkyd' exclusively to urethane oils as defined here. The urethane alkyds of this book are then called 'urethane-modified alkyds'.

The urethane linkage is resistant to alkalis and thus urethane oils and alkyds have better alkali resistance than ordinary alkyds. This is usually accompanied by better water resistance. Urethane oils and alkyds possess a property common to all polyurethane finishes: good abrasion resistance. It is claimed that they are superior to oils and alkyds for the dispersion of difficult organic pigments and carbon black, but in pigmented form they are prone to chalk and lose gloss earlier than normal alkyds and oleoresinous finishes.

The main outlets for urethane oils and alkyds are in varnishes for floors, boats and general use; in undercoats and in industrial maintenance finishes (where gloss retention is less important than long-lasting film integrity and resistance to water and chemicals).

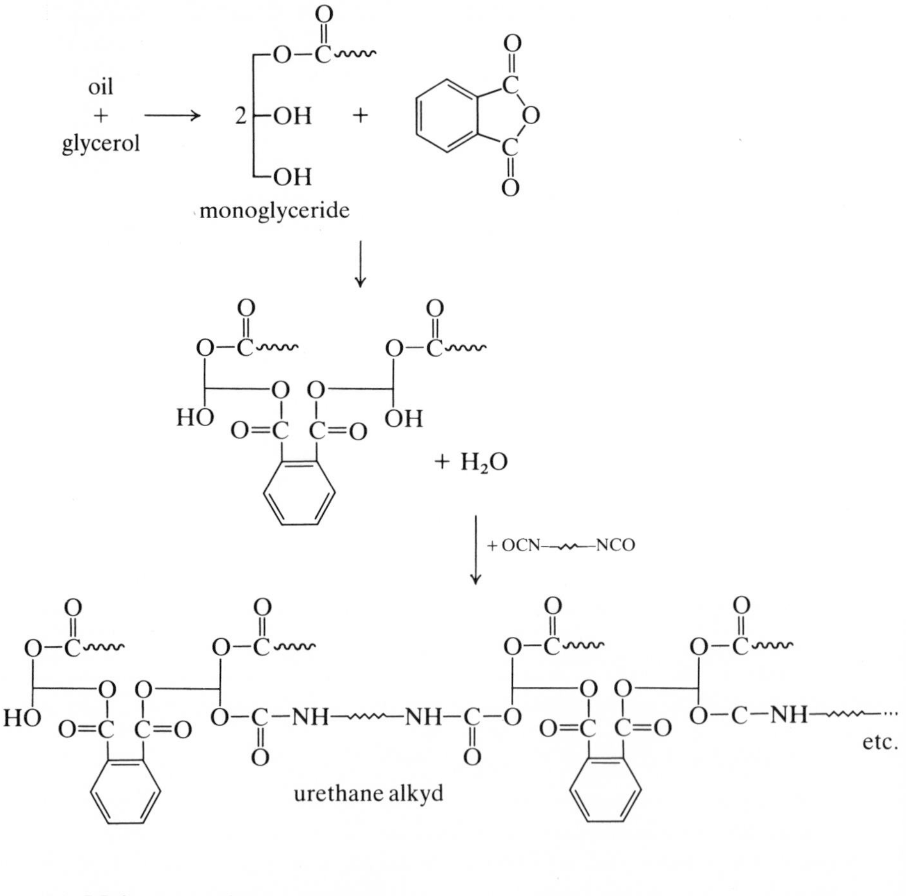

monoglyceride

urethane alkyd

### (b) Moisture-curing prepolymers

Finishes of this type do not form urethane linkages during the drying process. Like the first group of finishes, they dry by the second mechanism of Chapter 7, by reaction with the atmosphere. This time it is the water vapour in the air, not the oxygen, which reacts chemically with isocyanate in the finish. If a polyisocyanate is present, polymerization can occur:

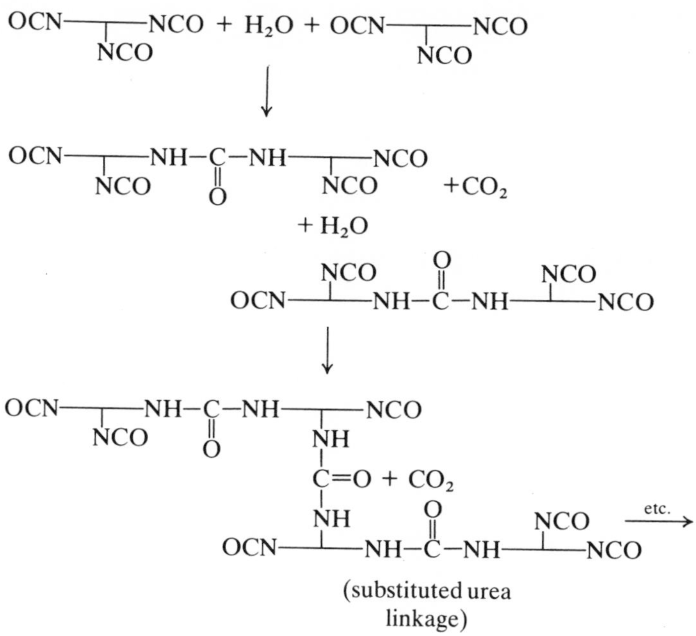

(substituted urea
linkage)

Drying is faster if the initial polyisocyanate molecules are large. Because of its wide usage in polyurethane foams, the cheapest diisocyanate available is the low molecular weight toluene diisocyanate (or TDI). This material is volatile, toxic and gives rise to marked yellowing in paint films. However, TDI can be converted into a larger molecule by reaction with a polyol:

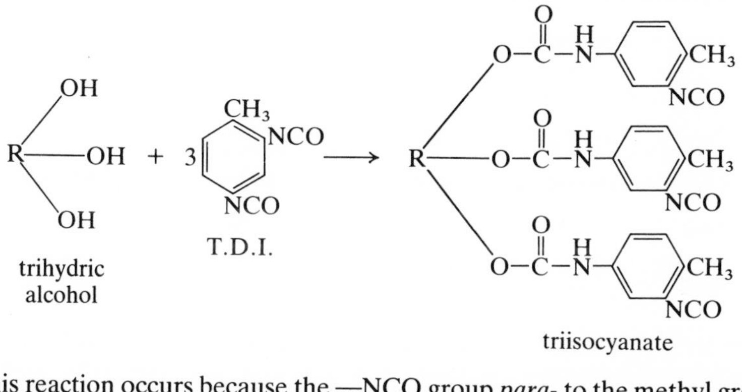

This reaction occurs because the —NCO group *para-* to the methyl group is several times more reactive than the one *ortho-* to it and will therefore react with the hydroxyl groups preferentially. Some *ortho-* groups will, of course, react and there is an added complication in that commercial TDI contains some of the 2,6-isomer. Thus side reactions leave some unreacted TDI in the final mixture, but methods of reducing this to a very low level have been devised and the resultant 'pre-polymer', as it is called, is safe for spraying with appropriate ventillation arrangements (see p. 210). Any fairly high molecular weight polyol can be used. Castor oil is such a material. Polyethers (see below) are often used.

By the route shown, or any other, a fairly high molecular weight poly-isocyanate with a minimum functionality of three (isocyanate groups) is produced. This is the film-former of the finish, reacting with water to give a cross-linked polymer film. The carbon dioxide produced in the process is formed slowly and diffuses through the film without causing 'bubbling' or 'popping'. The drying rate varies enormously with the humidity. Force-'drying' by steam is possible.

As the prepolymer reacts with moisture, this must be rigorously excluded from the can, since it will not form a skin on the paint but will eventually gel it right through. Another difficulty is the moisture found on most pigment surfaces. This must be removed, possibly by reaction with monoisocyanate, before the pre-polymer can be pigmented. If it is not, gelation occurs. Stability may be further enhanced by addition of moisture scavengers to the coating, such as ethyl orthoformate, $HC(\cdot O \cdot C_2H_5)_3$, or *molecular sieves*. The latter are sodium or potassium aluminium silicates, with a finely porous crystal structure. Only small molecules, such as water, can penetrate these fine pores and they are then strongly adsorbed within them. The main outlet for this type of finish – which has the typical polyurethane virtues of tough-ness and abrasion-resistance – is as a floor varnish.

### (c) Blocked isocyanate stoving finishes

At sufficiently high temperatures, the reaction products of isocyanates and certain other compounds decompose to reform the isocyanate:

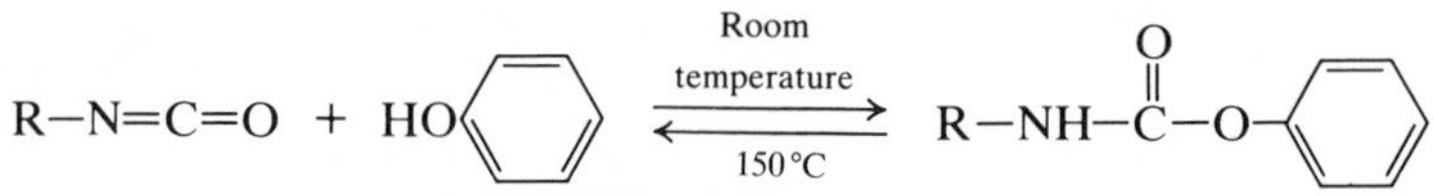

Hot phenol is an unpleasant by-product to release from a paint and polyisocyanates 'blocked' with phenols have not met with widespread use in coatings. However, the use of polyisocyanates blocked with ε-caprolactam is a popular route to *polyurethane powder coatings*.

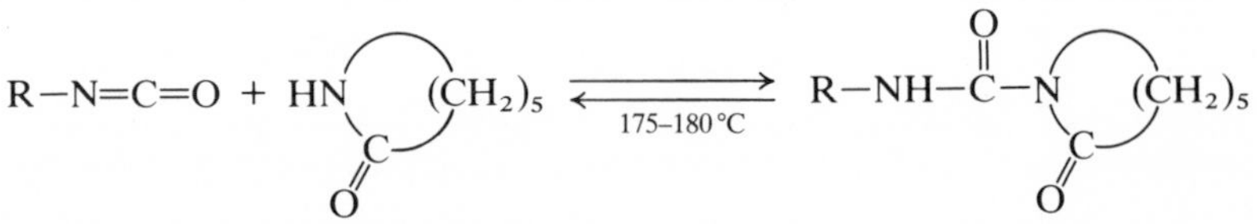

ε-caprolactam

The use of blocked isocyanates has proved an important method for cross-linking *cathodic electrodeposition coatings* (see p. 110). The resin in such a coating might be the reaction product of a bisphenol A diepoxide (with $n = 4$, see p. 184) and diethylamine:

$$
\begin{array}{c}
\text{N}\diagup\text{C}_2\text{H}_5 \\
\quad\diagdown\text{C}_2\text{H}_5 \\
\text{CH}_2-\text{CH}-\text{CH}_2\left[\text{O}-\text{C}_6\text{H}_4-\overset{\text{CH}_3}{\underset{\text{CH}_3}{\text{C}}}-\text{C}_6\text{H}_4-\text{O}-\text{CH}_2-\text{CH}-\text{CH}_2-\right]_4 \\
\quad\quad\text{OH}\qquad\qquad\qquad\qquad\qquad\qquad\qquad\qquad\text{OH}
\end{array}
$$

$$
-\text{O}-\text{C}_6\text{H}_4-\overset{\text{CH}_3}{\underset{\text{CH}_3}{\text{C}}}-\text{C}_6\text{H}_4-\text{O}-\text{CH}_2-\underset{\text{OH}}{\text{CH}}-\text{CH}_2-\text{N}\diagdown^{\text{C}_2\text{H}_5}_{\text{C}_2\text{H}_5}
$$

The hydroxyl groups in such a resin can be cross-linked by a blocked isocyanate such as

$$
\underset{\text{NCO}}{\overset{\text{CH}_3\;\text{NCO}}{\bigcirc}} \quad + \; 2\text{CH}_3\cdot(\text{CH}_2)_3\cdot\underset{}{\overset{\text{C}_2\text{H}_5}{\text{CH}}}\cdot\text{CH}_2\text{OH}
$$

175–190 °C
[+ dibutyl tin
dilaurate catalyst]   < 100 °C

$$
\underset{\text{NH}\cdot\text{CO}\cdot\text{O}\cdot\text{C}_8\text{H}_{17}}{\overset{\text{CH}_3\;\text{NH}\cdot\text{CO}\cdot\text{O}\cdot\text{C}_8\text{H}_{17}}{\bigcirc}}
$$

The two materials are mixed together and the tertiary amino groups in the modified epoxy resin are partially neutralized with acetic acid before emulsification in water.

## Two-pack paints

### (a) Activated prepolymer

This type of finish is an extension of the one-pack moisture-curing type. It is virtually the same finish, to which a small quantity of catalyst is added just before use. Tertiary amines, such as N,N-dimethylethanolamine $[\text{HO}\cdot\text{CH}_2\cdot\text{CH}_2\cdot\text{N}(\text{CH}_3)_2]$, are most frequently used as catalysts. Even if the moisture content of the air is low, the finish will harden rapidly, but the more reliable drying performance is obtained at the expense of a limited pot-life. Unfortunately, the activator is non-polymeric and so cannot be used for pigment dispersion. Pigments are dispersed in pre-polymer and must be free from moisture, as in the one-pack composition. The uncatalysed finish is water-sensitive, presenting a shelf-life storage problem as before.

### (b) Polyhydroxylic resin/isocyanate finish

In this finish the price of 'pot-life' is paid to gain the following advantages:

(1) Faster room-temperature drying than that of urethane oils and alkyds, by a mechanism independent of the atmosphere.
(2) Easier pigment dispersion in a resin free from isocyanate.
(3) Shelf-life stability of the paint, by isolation of the moisture-reactive component in an activator pack.
(4) Extreme formulating versatility, because of the wide variety of ingredients available for both packs.

The method is to place in the 'paint' pack, a solution of a linear or branched polymer containing hydroxyl groups, plus suitable additives. The activator pack contains a solution of a moderate molecular weight poly-isocyanate. This may be made by the prepolymer route, but usually from a lower molecular weight polyol. A suitable polyol might be trimethylol propane. Since the hydroxyl spacing in the resin can be made large, it is not necessary to have widely spaced isocyanate groups to get flexibility in the cross-linked film. Another polyisocyanate in commercial use is diphenyl-methane-4,4'-diisocyanate (MDI). Polyisocyanurates can be used. These are produced when isocyanates react with themselves at high temperatures.

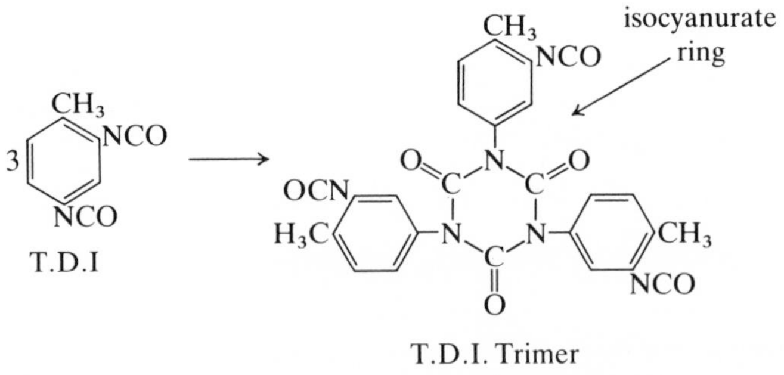

T.D.I. Trimer

In the equation a trimer is shown, but more complicated ring structures will also be formed, due to reaction of some of the three unreacted trimer —NCO groups with further TDI.

The resin, as has been stated, may be any resin containing hydroxyl groups. This includes castor oil, alkyds, nitrogen resins, epoxy resins, cellulose derivatives and so on, provided always that, if a mixture is used, then the ingredients must be compatible with one another. However, the most popular resins used are *saturated polyesters, acrylic resins* and, to a lesser extent, *polyethers.*

*Polyethers* are made by reacting ethylene oxide or propylene oxide with a polyol in the presence of an acid ($BF_3$) or basic ($NaOH$) catalyst, e.g.:

$$
\begin{array}{c}
CH_2OH \\
|\\
CHOH \\
|\\
(CH_2)_3 \\
|\\
CH_2OH \\
\text{1,2,6-hexane} \\
\text{triol}
\end{array}
\quad + \quad 3n \quad
\begin{array}{c}
CH_2\!-\!CH\!-\!CH_3 \\
\diagdown\!\!O\!\!\diagup \\
\text{propylene} \\
\text{oxide}
\end{array}
\quad \xrightarrow[\substack{120-200\,°C \\ \text{catalyst}}]{\text{pressure}} \quad
\begin{array}{c}
CH_3 \\
|\\
CH_2\cdot(O\cdot CH\cdot CH_2)_n\cdot OH \\
CH_3 \\
|\\
CH\cdot(O\cdot CH\cdot CH_2)_n\cdot OH \\
|\\
(CH_2)_3 \quad CH_3 \\
|\quad\quad | \\
CH_2\cdot(O\cdot CH\cdot CH_2)_n\cdot OH \\
\text{polyether}
\end{array}
$$

It can be seen that the polymer chains contain multiple alkali-resistant ether linkages and end in hydroxyl groups. Thus the number of hydroxyl groups in the polymer is the number of hydroxyl groups in the polyol starting ingredient. Molecular weights vary from 400 to 4000.

Although more expensive, *saturated polyester resins* have better colour retention and water resistance. Ether linkages are water-sensitive (diethyl ether, $C_2H_5\cdot O\cdot C_2H_5$, is partly soluble in water) and in polyethers they are repeated throughout the polymer chain without the separation by long, water insensitive hydrocarbon chains which could offset their effect.

Saturated polyester resins have been defined in Chapter 12 and their use with nitrogen resins was described in Chapter 13. The resins used in polyurethane finishes might contain a selection from the ingredients list on p. 167 and, amongst others:

| | |
|---|---|
| 1,3-butane diol | $HO\cdot CH_2\cdot CH_2\cdot CHOH\cdot CH_3$ |
| 1,4-butane diol | $HO\cdot CH_2\cdot CH_2\cdot CH_2\cdot CH_2\cdot OH$ |
| 1,6-hexane diol | $HO\cdot CH_2\cdot CH_2\cdot CH_2\cdot CH_2\cdot CH_2\cdot CH_2\cdot OH$ |
| 1,2,6-hexane triol | $HO\cdot CH_2\cdot CHOH\cdot CH_2\cdot CH_2\cdot CH_2\cdot CH_2\cdot OH$ |

Again hydroxyl groups are to be found at the ends of chains. The number present is greater than that likely to be found in an ordinary saturated polyester, because the excess of hydroxyl groups over carboxyl in the resin ingredients is much higher than that usually allowed. For this reason also, some unreacted hydroxyls from triols are likely to be found in the middle of polymer chains:

$$\cdots—O\cdot CO\text{—}\sim\sim\text{—}CO\cdot O\text{——}\underset{\underset{CH_2OH}{|}}{}O\cdot CO\text{—}\sim\sim\text{—}CO\cdot O—OH$$

(only ester links and hydroxyl groups shown)

The formulation of *hydroxyl-containing acrylic copolymers* has already been discussed on p. 141 and their use with nitrogen resins on p. 178. More recently, hard, extremely flexible and durable coatings have been formulated by using these acrylic resins with aliphatic polyisocyanates (see *Yellowing* below). Such coatings have been used for refinishing motor cars, where their rapid cure at room temperature is advantageous, and they have also been proposed for the factory finishing of motor cars at high solids and for coating hardboard sheet to produce an artificial ceramic tile appearance. They are expensive by acrylic standards, so cost is minimized by formulating with acrylic resins of low hydroxyl functionality, so that only small amounts of the costly isocyanate are needed. Inevitably, low cross-link densities give a chemical resistance which is inadequate where requirements are exacting.

The design of the resin containing hydroxyl groups is important to film properties, since the spacing of hydroxyl groups along the chain will influence the flexibility. Close spacing will mean frequent cross-links and poor flexibility. Pronounced branching in the resin also tends to the same result. Dense cross-linking is usually accompanied by increased hardness and improved resistance to water, solvents and chemicals.

From this emphasis on hydroxyl groups, it might be thought that the hardening process consists solely of urethane formation between resin and polyisocyanate. This is obviously not so, since water vapour is usually present and the isocyanate reactions with water must proceed in competition with the reaction with hydroxyl. This will lead to consumption of isocyanate groups and linking of isocyanate molecules by urea linkages. With a difunctional isocyanate this tends to space the resin chains farther apart, giving increased flexibility:

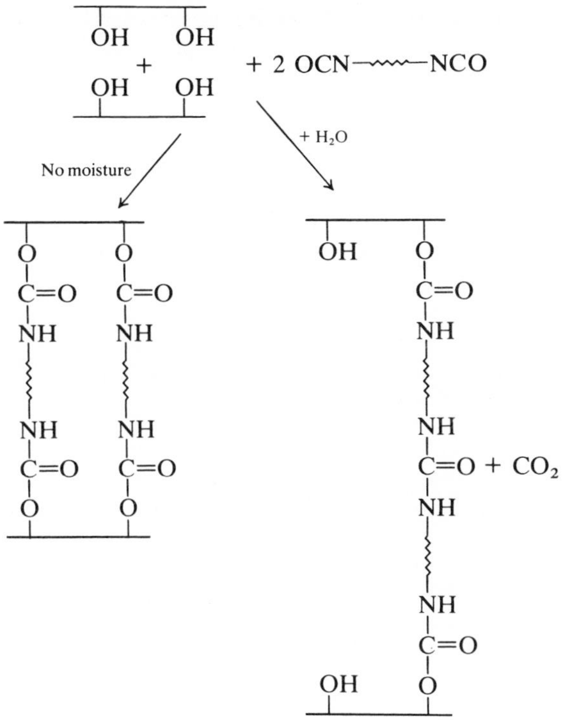

But if the functionality is greater than two, branching occurs and increased overall isocyanate functionality results:

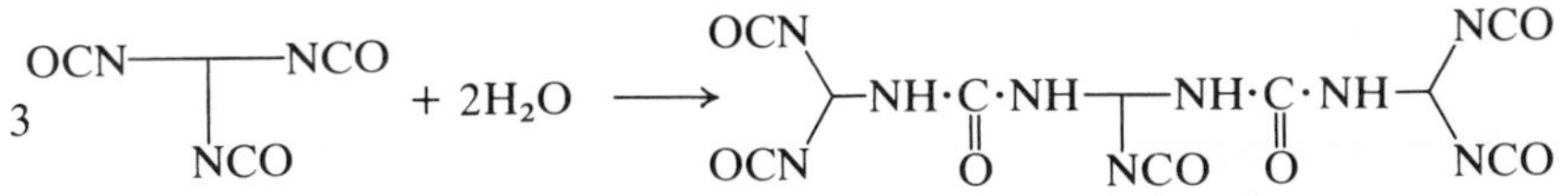

(functionality 3)                    (functionality 5)

To match the number of available resin hydroxyl groups in the paint pack with the same number of isocyanate groups in the activator pack, is therefore not practical and it is usual to have a ratio of —NCO/—OH of $1 \cdot 1/1$ to $1 \cdot 3/1$, allowing for reaction with moisture. Increased flexibility is gained by reducing the ratio; greater hardness and chemical resistance result from increasing it, together with faster curing.

In addition to the uses already mentioned, polyol/isocyanate two pack finishes have wide applications for furniture, floors and boats; as corrosion-resistant metal finishes and for coating plastic, rubber, concrete and masonry. Exterior durability is good if ingredients are correctly chosen and paints well formulated.

### (c) Vapour curing

The reactions of isocyanates in both prepolymer and polyhydroxylic resin/ isocyanate finishes are catalysed by various catalysts, including tertiary amines. If the catalyst is added to the coating, fast cure times at room temperatures and relatively short pot lives result. However, if the amine is withheld from the coating and introduced as a vapour, either by vapouriz- ation into the first stage of a curing chamber, or by injection into the atomizing air of a spray gun or the directing air of an air-assisted electrostatic spray gun, then the coating is cured rapidly, catalysed by permeation of the amine into it. The pot life of the polyhydroxylic resin/isocyanate two pack coating is considerably lengthened, while the shelf life of the prepolymer coating is long if moisture can be excluded successfully.

N,N-dimethylethanolamine is one of the preferred amines, because its combination of good catalytic activity, adequate volatility and relatively low toxicity is an effective compromise. Amines are not equally effective catalysts for all types of isocyanate and care must be taken about selecting aliphatic isocyanates in particular (see p. 210) to achieve best cure rates. Vapour curing has been shown to work successfully in small industrial painting installations and has potential benefits on metal castings, which absorb heat intended for curing, and on some plastic substrates (e.g. GRP), where surface defects can cause blow holes in the paint coating during stoving.

### (d) Solventless finishes

Since liquid polyols (e.g. castor oil, polyethers, low molecular weight polyesters) and liquid isocyanates (e.g. diphenylmethane diisocyanate) are available, it is theoretically possible to produce 100 per cent solids finishes. However, carbon dioxide has great difficulty in diffusing through thick films of such viscous coatings and bubbles of the gas form holes and craters. The coatings will only be successful if moisture from all sources can be reacted or absorbed harmlessly. The use of molecular sieves for this purpose has been recommended. Coatings with pot-lives of 30–60 minutes are applied by manual spreading with various implements on concrete floors and road surfaces. Twin-feed spray guns may also be used.

## General matters

One or two matters common to polyurethane finishes of all types must now be covered briefly.

## Solvents

Except for urethane oil and alkyd finishes, all polyurethanes require solvents that will not react with isocyanates: alcohols and ether-alcohols are obviously out. The water content of the solvents should be negligible. Manufacturers often label their solvents 'urethane grade', implying low water content. Solvents readily miscible with water, e.g. acetone, should be treated with suspicion, as they can take up water from the atmosphere. Obviously the exact choice of solvent will vary considerably with the finish composition.

## Additives

The usual thickeners and flow agents can be used. The usual driers and antioxidants are used in urethane oils and alkyds. Catalysts which speed up the curing reaction (and reduce pot-life) are tertiary amines and metal salts, particularly tin salts. The latter should be checked for toxicity before inclusion in paint.

## General properties

In spite of the wide variety of types of polyurethane finish, certain characteristics are common to the family as a whole. These are toughness and abrasion resistance, combined with flexibility, good chemical resistance and good adhesion.

An outstanding feature of paints that dry by reaction of isocyanate, is that hardening will occur even at 0 °C. The finishes can therefore be used in conditions unsuitable for epoxies and acid-catalysed nitrogen resin finishes.

## Yellowing

In Chapter 6 it was shown that colour is associated with certain chemical groups and that many of these contained nitrogen. Thus any film-former containing nitrogen atoms should be suspect for colour retention, until it has been shown to be otherwise. It is not a matter of whether chromophoric groups are present in the freshly formed polymer, but whether they are likely to be produced as decomposition products on exposure to air, light, moisture and warmth. Polyurethanes based on aromatic isocyanates are prone to yellowing (just as epoxies based on aromatic amines discolour, Chapter 14). The effect of aromatic rings on colour is covered in Chapter 6. Prepolymers made from aliphatic isocyanates do not yellow, but are much less reactive than their aromatic counterparts. This cure rate problem is overcome by reacting three molecules of aliphatic diisocyanate with one of water, to produce a triisocyanate containing urea linkages (together with more complex molecules not shown in the following idealized equation):

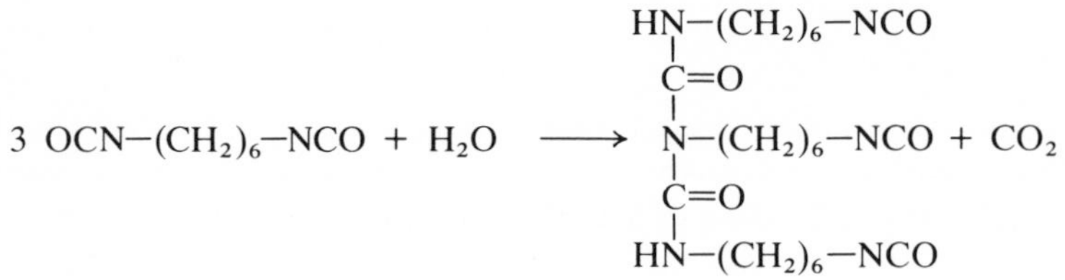

$$
3 \; OCN-(CH_2)_6-NCO + H_2O \longrightarrow \begin{array}{l} HN-(CH_2)_6-NCO \\ \quad | \\ \quad C=O \\ \quad | \\ N-(CH_2)_6-NCO \\ \quad | \\ \quad C=O \\ \quad | \\ HN-(CH_2)_6-NCO \end{array} + CO_2
$$

Other aliphatic diisocyanates of interest are:

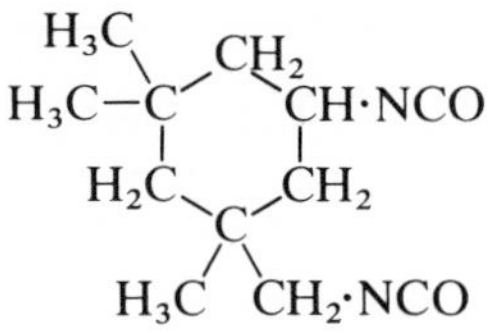

isophorone diisocyanate
(IPDI)

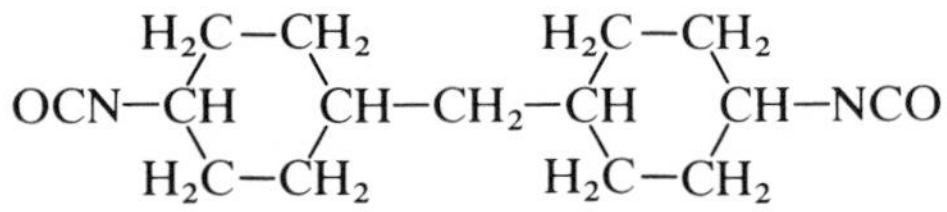

dicyclohexyl methane-4,4'-diisocyanate
(H$_{12}$MDI)

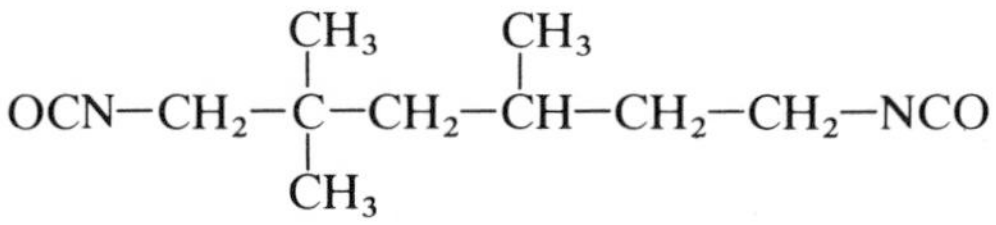

$$ OCN-CH_2-\underset{\underset{CH_3}{|}}{\overset{\overset{CH_3}{|}}{C}}-CH_2-\underset{\underset{}{|}}{\overset{\overset{CH_3}{|}}{CH}}-CH_2-CH_2-NCO $$

2,2,4-trimethyl-1,6-hexane diisocyanate
(TMDI)

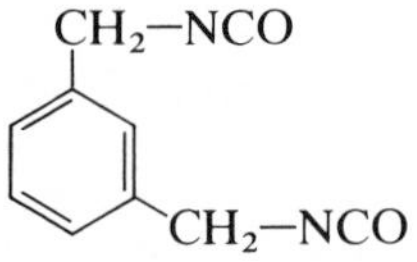

xylene diisocyanate
(XDI)

Formulations based on non-yellowing aliphatic isocyanates with good durability outdoors are now well established, but cost significantly more than corresponding products based on aromatic isocyanates.

**Toxicity**

Volatile isocyanates, such as TDI and hexamethylene diisocyanate, are severe irritants to the eyes, nose and throat. They are also respiratory sensitizers, causing asthmatic symptoms in sensitized individuals. Higher molecular weight polyisocyanates for paints therefore should contain very low levels of these isocyanates, so that the isocyanate content of the air around them remains below the OEL or TLV (see p. 125). Even if they contain only low levels of volatile isocyanates, precautions must be taken to prevent inhalation of spray droplets containing unreacted isocyanates (either excellent extraction of overspray in the spray booth, or the wearing of fresh air hoods by those in the vicinity). The uncured liquid finishes can also be skin irritants.

Finishes based on urethane oils and alkyds and blocked isocyanates do not present the same hazards.

Some examples of polyurethane formulae follow.

---

Oil-modified polyurethane white decorative gloss finish

| | wt % | |
|---|---|---|
| Rutile titanium dioxide | 28·7 | **Pigment** |
| Zinc oxide | 1·5 | |
| Polyurethane 2388 | 65·5 | **Resin solution** |
| Cargill N lecithin | 0·3 | **Pigment dispersant** |
| Antioxidant | 0·1 | To improve stability in the can |
| Cobalt naphthenate (6% Co) | 0·1 | |
| Calcium naphthenate (5% Ca) | 0·3 | **Driers** |
| Zirconium complex (6% Zr) | 0·3 | |
| Mineral spirits | 3·2 | **Solvent** |
| | 100.0 | |

---

Ready for brushing at 63% solids.
*Polyurethane 2388* is a urethane oil/alkyd based on safflower oil at 50% solids in mineral spirits (white spirit) and is made by Cargill Inc., USA.
*Cargill N lecithin* is soya lecithin.

---

Moisture-cure prepolymer floor finish

| | wt % | |
|---|---|---|
| Polyurethane 3651 | 66·4 | **Film former** |
| Xylol | 33·2 | **Solvent** |
| Dimethylethanolamine* | 0·4 | **Catalyst*** |
| | 100·0 | |

---

Ready for application to wood or concrete floors at 40% solids.
*Polyurethane 3651* is a moisture-curing prepolymer at 60% solids in xylol and 'Cellosolve' acetate and is made by Cargill Inc., USA.
* Catalyst may be added just before use *to speed up cure at low humidities* (when a pot life of 8–24 hours is obtained) *or it may be omitted altogether.*

Two-pack flexible coating for integral skin polyurethane foam

| | wt % | |
|---|---|---|
| **(a) *Paint*** | | |
| Cadmopurbordo N | 5·25 ⎫ | |
| Thiofast Red MV 6606 | 1·75 ⎭ | **Pigment** |
| Silica OK 412 | 1·72 | **Matting extender** |
| Desmophen 1200 | 6·66 ⎫ | |
| Desmophen 1800 | 13·32 ⎬ | **Resins** |
| Vinylite VAGH | 1·97 ⎭ | |
| 'Cellosolve' acetate | 24·86 ⎫ | |
| Butyl acetate | 6·08 ⎪ | **Solvents** |
| Methyl ethyl ketone | 32·25 ⎬ | |
| Toluene | 6·08 ⎭ | |
| Desmorapid PP | 0·06 | **Catalyst** |
| | 100·0 | |
| **(b) *Activator*** | | |
| Desmodur HL60 | | **Polyisocyanate** |

Mixing ratio: paint/activator, 6·56/1 by weight.
Spray application solids: 34%.
Pot life (closed container): approx. 70 h.
*Desmophen 1200 and 1800* are slightly branched polyesters from Bayer.
*Vinylite VAGH* is a copolymer of vinyl chloride and other monomers from Bakelite.
*Desmodur HL60* is a polyisocyanate of mixed aromatic/aliphatic type at 60% solids in ester solvent from Bayer.

# Sixteen

# Unsaturated polyesters and acrylics

In Chapter 12, we mentioned that the departure from the 100 per cent solids of the old oil paints, to the lower solids of modern finishes, was a consequence of the demand for paints with better performance. The paint industry's determination to provide the improved properties at higher solids was also mentioned. We have already seen in Chapter 14 how the epoxy/polyamine system meets these requirements. The finishes under discussion in this chapter are another approach to the problem.

A 100 per cent solids paint is achieved when the film-forming ingredients are fluid enough for application *and are substantially involatile.* Since the polyester portion of polyester/glass fibre mouldings fulfils the first part of these conditions and goes some way towards satisfying the second part, it is clearly a potential paint material. The development of the plastics material into a practical polyester finishing system for wood was completed first in Germany in the 1950s. The result was a smooth, hard, glass-like finish with excellent resistance to most forms of damage that occur in domestic usage. Polyesters have, however, other features which have limited their exploitation over other substrates and in other paint markets.

The development in the 1960s of extremely rapid methods of curing these coatings with ultra-violet and electron beam radiation, has led to an extension of the chemistry to compositions based on resins and monomers containing acrylic unsaturation. These coatings will also be discussed in this chapter.

## Ingredients of unsaturated polyester coatings

First comes the *unsaturated polyester resin* itself. From Chapters 12 and 15 we should have a clear idea of what is meant by a polyester resin and, in particular, a saturated polyester resin. Although a drying-oil alkyd is, in a sense, an unsaturated polyester resin, the term has come to be applied solely to polyester resins based on components which introduce unsaturation *directly into the polyester backbone.* This unsaturation must be capable of direct addition copolymerization with vinyl monomers. To give a linear polymer, any of the dibasic acids or dihydric alcohols mentioned in Chapters 12 and 15 may be used, but the resin should include some unsaturated components. These are usually, though not necessarily, acids, e.g.:

maleic anhydride,

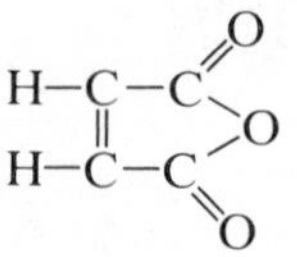

$$\begin{array}{c} H-C-C \diagdown O \\ \| \quad \diagup O \\ H-C-C \diagup \\ \diagdown O \end{array}$$

citraconic anhydride,

$$\begin{array}{c} CH_3 \\ | \\ C-C \diagdown O \\ \| \quad \diagup O \\ HC-C \diagup \\ \diagdown O \end{array}$$

fumaric acid,

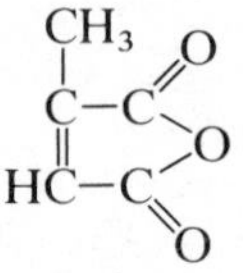

$$\begin{array}{c} H-C-COOH \\ \| \\ HOOC-C-H \end{array}$$

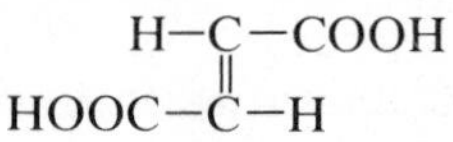

itaconic acid,

$$\begin{array}{c} CH_2 \\ \| \\ C-COOH \\ | \\ CH_2 \cdot COOH \end{array}$$

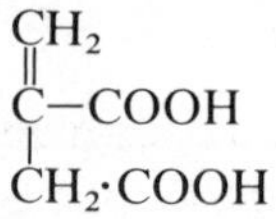

Part of a typical unsaturated polyester chain might be:

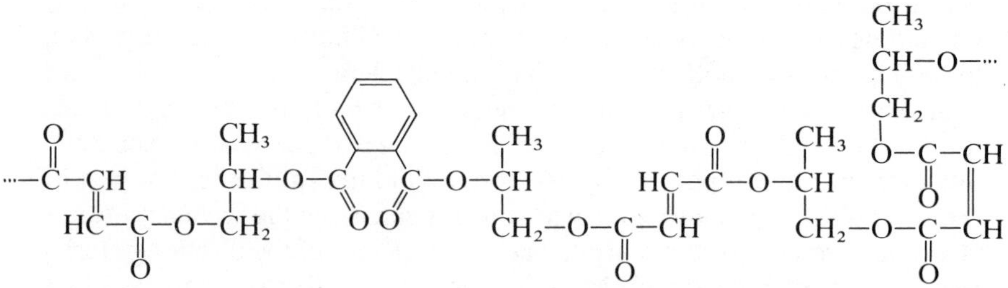

$$\cdots-\overset{O}{\underset{\|}{C}}-\underset{\underset{\|}{HC}-\underset{\|}{C}-O-CH_2}{CH} \quad \underset{\overset{CH_3}{|}}{CH-O-C} \quad \overset{\diagup}{\underset{O\ O}{\diagdown}} C-O-\underset{\underset{\|}{CH_2-O-C}}{\overset{CH_3}{\underset{|}{CH}}} \quad \underset{O}{HC-\overset{O}{\underset{\|}{C}}-O-CH} \quad \cdots$$

This would be made from 1,2 propylene glycol, phthalic anhydride and (probably) maleic anhydride (since the maleic form of the rigid double bond,

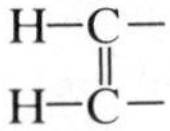

$$\begin{array}{c} H-C- \\ \| \\ H-C- \end{array}$$

largely changes to the fumaric form,

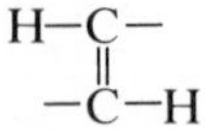

$$\begin{array}{c} H-C- \\ \| \\ -C-H \end{array}$$

at the temperature of the resin cook). Typical molecular proportions might vary from 22:10:10 to 22:5:15 (a slight excess of hydroxyl in each case). Normal alkyd methods of preparation are used, with, of course, inert gas over the liquid to prevent oxidation of the unsaturation. Since there is no monoglyceride stage, all the ingredients are usually present from the start of the reaction. If *iso*-phthalic acid is used, it is reacted with the diol before the maleic anhydride is added. If this is not done, unreacted insoluble *iso*-phthalic acid will remain at the end of the resin preparation.

The second ingredient is the *copolymerizable vinyl monomer*. Since the polyester resin is highly viscous, the monomer must be extremely fluid and a solvent for the resin, so that the combined solution is fluid enough for application. When the resin is cooling, but still fluid, the monomer is introduced to produce the resin solution, which is the basis of the finish. In spite of the reactivity of the two components, no reaction can begin unless free radicals are produced in the solution. These radicals can only be produced by:

(1) Ultra-violet light. This is excluded by storage in tins.
(2) Heat. Normal storage temperatures are quite safe.
(3) The presence of an unstable compound. Such a compound may be present in trace quantities, possibly due to (4).
(4) Oxygen attack on double bonds or peroxide formation at reactive carbon atoms. Oxygen is never completely excluded from cans though, as we shall see later, its presence is less likely to cause polymerization with this system than with drying oils.

Since there is some chance of polymerization and gelling due to (3) and (4), both the monomer itself and the polyester resin solution contain small quantities of *inhibitor*. Hydroquinone,

$$HO \langle \rangle OH,$$

and other phenolic materials are effective in reacting with free radicals to destroy their free radical nature, e.g.:

$$RO_2 \cdot + X \langle \rangle OH \longrightarrow RO_2H + X \langle \rangle - O \cdot \xrightarrow{+RO_2 \cdot} X \langle \rangle = O$$

(where X is an alkyl group and the inhibitor a trialkyl phenol.)

Quantities of less than 0·05 per cent inhibitor in the resin solution will give stability for several months, provided conditions (1) and (2) above are avoided. Further quantities may be added to the paint itself to delay curing. A total of up to about 0·2 per cent might be included.

Any relatively involatile monomer with the vinyl ($CH_2{=}CH{-}$) or vinylidene ($CH_2{=}C{<}$) grouping present is theoretically suitable. In practice, many are not sufficiently reactive. Of the remainder, styrene is very reactive and cheap. It is usually the chief (or only) monomer, but may have a comonomer such as vinyl toluene, methyl methacrylate, ethylene glycol dimethacrylate, triallyl cyanurate or diallyl phthalate.

$$CH_2 \cdot O \cdot CO \cdot \underset{\underset{CH_3}{|}}{C} {=} CH_2$$
$$CH_2 \cdot O \cdot CO \cdot \underset{\underset{CH_3}{|}}{C} {=} CH_2$$

ethylene glycol dimethacrylate

diallyl phthalate

triallyl cyanurate

The solution, having been made stable in the can, must now be made reactive on the coated surface. This is done by introducing factor (3) in relatively large quantities. The substance chosen to decompose and produce free radicals is an *organic peroxide*. A solution of this material in unreactive solvent forms the activator pack of the two-pack finish. The amount of solvent is chosen to give a suitable finish/activator mixing ratio, e.g. 10/1 by volume. These peroxides are all relations of hydrogen peroxide and can be subdivided into five types. All contain at least one peroxide (—O—O—) unit (see p. 217).

The ketone, diacyl and hydroperoxide types are most frequently used. Commercial ketone peroxides are not pure and contain substantial amounts of hydroperoxide.

If the finish is to be stoved, a wide range of peroxides is suitable, depending on the temperature to be used and the pot-life required. Lauroyl, 2,4-dichlorobenzoyl and benzoyl peroxides (acyl peroxides); methyl ethyl ketone (MEK) peroxide and cyclohexanone peroxide (ketone peroxides) and cumene hydroperoxide are examples.

However, if the finish is to be cured at room temperature, there is no peroxide which can, on the one hand, remain relatively stable in the activator pack and, on the other, produce free radicals fast enough to give rapid hardening when activator and finish are mixed. Peroxide decomposition is accelerated by the addition of a metal salt from the drier range. The mechanism of the catalysis is as given in Chapter 12 for hydroperoxides. The combination of cobalt and either of the above ketone peroxides is widely used. Another combination is benzoyl peroxide and a tertiary amine, but this has the drawback that the films produced are prone to yellowing.

Since the *accelerator* acts only on the peroxide, it is safely incorporated in the finish pack, though it is often added to pigmented compositions just before use, to avoid slow deactivation during storage, by adsorption on the

ORGANIC PEROXIDES

| Attached group(s) | Peroxide type | Comments | Example |
|---|---|---|---|
| 1. $2R-\overset{\overset{\displaystyle O}{\|}}{C}-$ <br><br> acyl groups | $R-\overset{\overset{\displaystyle O}{\|}}{C}-O-O-\overset{\overset{\displaystyle O}{\|}}{C}-R$ <br><br> diacyl peroxide | R may be aliphatic <br> or <br> aromatic | 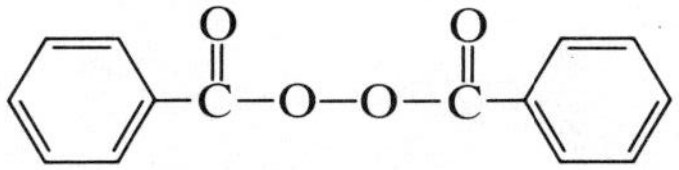 <br> benzoyl peroxide |
| 2. $R-\overset{\overset{\displaystyle O}{\|}}{C}-$ and $-R'$ <br><br> alkyl | $R-\overset{\overset{\displaystyle O}{\|}}{C}-O-O-R'$ <br><br> peracid ester | ester of the peracid, <br><br> $R-\overset{\overset{\displaystyle O}{\|}}{C}-O-O-H$ | 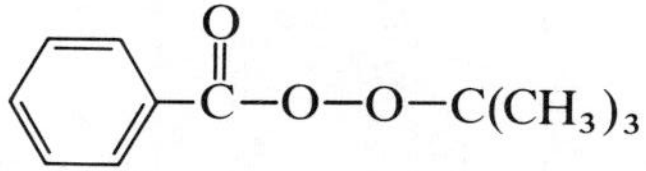 <br> $t$-butyl perbenzoate |
| 3. $\overset{\displaystyle R}{\underset{\displaystyle R'}{>}}C<$ | 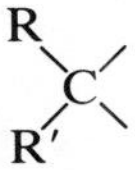 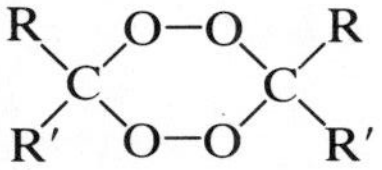 $\overset{R}{\underset{R'}{>}}\overset{O-O}{\underset{O-O}{C}}<\overset{R}{\underset{R'}{}}$ <br><br> ketone peroxide | or aldehyde peroxide <br> if R' is H. | 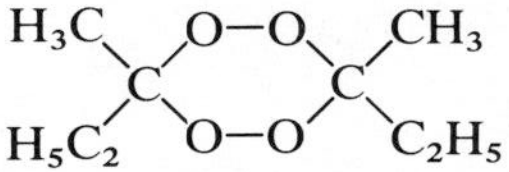 methyl ethyl ketone peroxide <br> |
| 4. $R-$ and $-H$ | $R-O-O-H$ <br><br> hydroperoxide | | 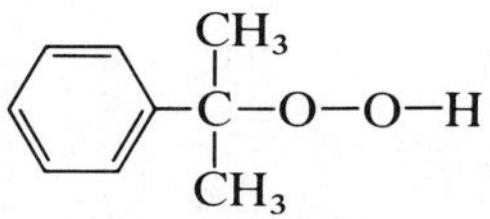 cumene hydroperoxide <br> |
| 5. $2R-$ | $R-O-O-R$ <br><br> dialkyl peroxide | | $(CH_3)_3{\cdot}C-O-O-C{\cdot}(CH_3)_3$ <br> di $t$-butyl peroxide |

particle surfaces. On no account should it be mixed with the peroxide directly, since the combination is explosive. Many pure peroxides explode if subjected to sudden shock or heat. They are therefore supplied in a more stable form as pastes, or as solutions in a high boiling liquid (e.g. the plasticizer dibutyl phthalate).

We have now covered the main finish ingredients which may be summarized thus:

| *Finish pack* | *Activator pack* |
|---|---|
| Pigment (if required) | Peroxide |
| Polyester resin | Solvent (to adjust mixing ratio) |
| Monomer | |
| Inhibitor | |
| Accelerator (if required) | |

Thick films (125–250 $\mu$m) can be applied to verticals in one coat without sagging, by the inclusion of the usual thickening additives. The resin solution is easily pigmented, but care must be taken to avoid overheating (leading to gelation) in the milling process.

## Pot-life and methods of activation

If slow hardening is acceptable, or if heat can be used to cure the finish, the simple two-pack arrangement described above will give pot-lives long enough to allow the user to finish a large number of articles with one batch of activated finish. But the polyester finishes are capable of really rapid hardening at room temperature. Touch-dry times of 15–30 minutes are not uncommon and mechanical sanding and polishing can be done after 4 hours. This sort of speed is very attractive to mass production furniture makers. To get this hardening rate, the peroxide and accelerator levels must be adjusted so that the pot-life of the above mixture is only *10–20 minutes*. The finish must be activated in small separate batches and spray-guns must be rinsed out scrupulously after each batch has been used, to prevent blockages due to gelled finish. To avoid the drawbacks of this 'pot-mix' system, several alternative activation methods have been devised:

### (a) Dual feed

In this method, the finish and activator are metered mechanically along separate feed lines and mix only at the instant of application, usually after both materials have left the application equipment. Special twin-feed spray-guns have to be used. The main objections are that sometimes mixing is not efficient enough and faults in the mechanical metering device can lead to insufficient or excess peroxide. If mechanical failure occurs, the defect will probably remain undetected until several objects have been coated and probably ruined. However, the latest equipment is quite good and the method is particularly suited to electrostatic spraying.

### (b) Wet-on-wet or 'sandwich' process

A longer pot-life of several hours can be obtained if the finish contains peroxide but no accelerator. In this process, such a finish is applied as the first coat, and is immediately followed by a second coat of finish which contains no peroxide, but has a high dose of accelerator. Free radicals are produced in high concentration at the inter-coat boundary, from whence polymer chains grow outwards into the two coats. The process is particularly suited to curtain-coating. Curtains of the two finishes fall continuously from slits in two 'heads' placed one behind the other and the object passes through one curtain, then the other, in one pass. One drawback of the process is that the two coats have to be carefully matched and the correct film thicknesses applied or serious defects in appearance and curing can occur. Another is that the pot-life problem is not avoided; it is made a little more tolerable.

### (c) Contact process

In this process the peroxide is added to a simple, ultra-low solids, non-reactive lacquer. A thin coat of lacquer is applied. This is followed by polyester finish containing accelerator. Free radicals again form at the inter-coat boundary and curing will proceed, even through a second coat of polyester applied several minutes after the first. This process completely avoids the pot-life problem – a big gain – but it has several disadvantages. Delayed solvent evaporation from the lacquer (or basecoat) can cause the entire finish to 'sink' after a few weeks ageing, showing up wood grain pattern, for example, in what was previously a smooth surface. In furniture finishing, special peroxide-resistant dyes must be used in the stains applied under the basecoat. If an appreciable area is 'missed' by the basecoat, either curing will not occur in that area, or the area will be comparatively soft. Finally, the basecoat dry film *must* be thin to minimize sinkage. Therefore basecoat solids must be low and a lot of wasted solvent must be paid for by the customer.

Nevertheless, this is the most widely preferred of all methods. In these days, when so much furniture is made from pre-finished flat boards, the basecoat can be applied accurately in thin coats by roller coater and thus can contain higher solids. The finish is then applied by a curtain coater, which is positioned immediately after the roller coater on the production line.

## Air-inhibition

Whichever method of activation is used, decomposition of peroxide leads to free radical attack, probably on the styrene:

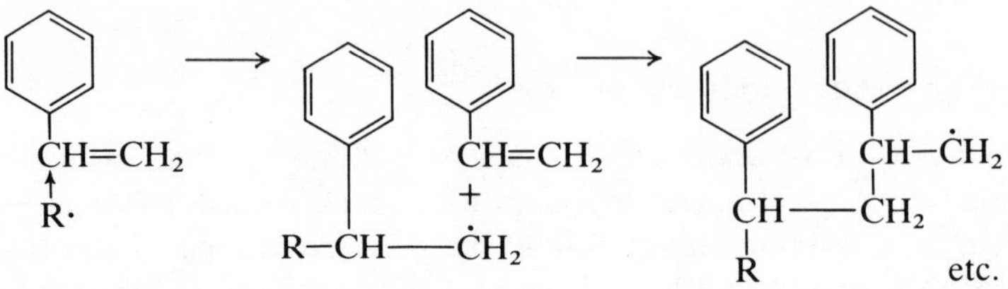

After quite a short polystyrene chain is formed, the free radical end encounters polyester unsaturation and copolymerization occurs:

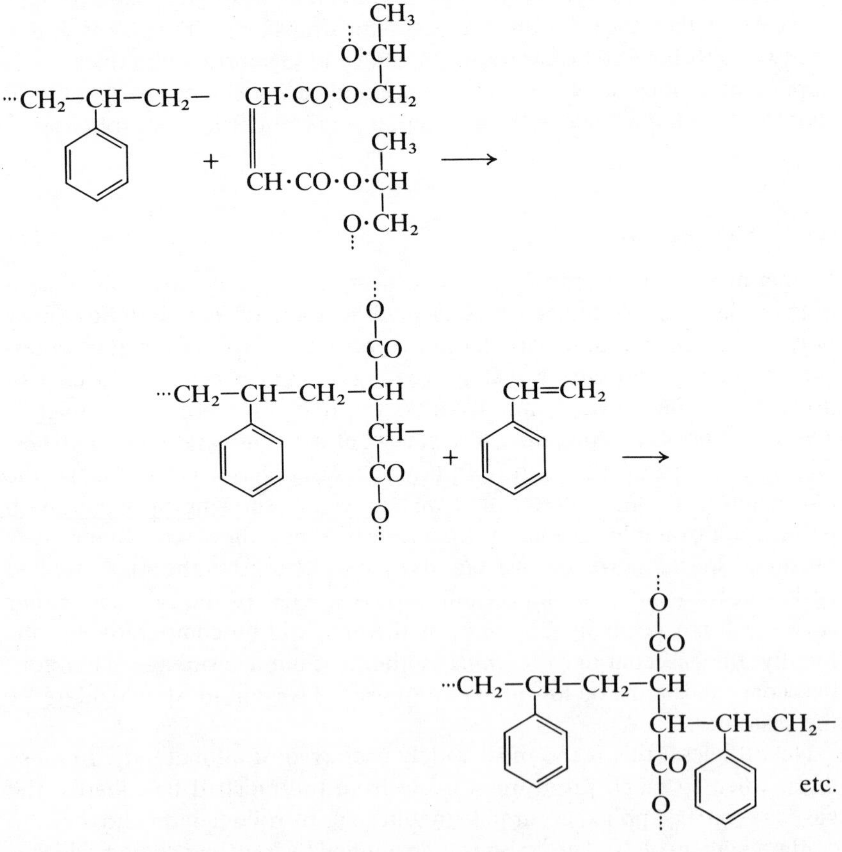

A simplified picture of the resulting cross-linked polymer is given in Fig. 33.

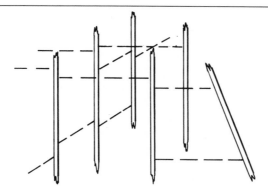

**Fig. 33** The cross-linking of an unsaturated polyester finish. The broken rods symbolize portions of the polyester chains and the dashes represent styrene molecules.

However, at the surface of the paint a different story unfolds:

As explained in Chapter 12, polyperoxides are extremely stable and styrene polyperoxides will not initiate the cross-linking copolymerization. They decompose eventually to aldehydes. Unreacted styrene (boiling point, 146 °C) evaporates from the surface, leaving only the sticky polyester resin and styrene polyperoxides behind. The surface will never harden.

There are at present two practical solutions to this problem. One is to include a very small proportion of paraffin wax, which is soluble in styrene, but becomes insoluble as the finish cures. It separates to the surface of the film, forming a continuous layer which (*a*) prevents oxygen from reacting with the styrene free radicals and (*b*) prevents evaporation of styrene. The wax gives a matt surface which is acceptable for lower gloss finishes. If a higher gloss finish is required, it must be removed by mechanical sanding, followed by mechanical polishing to a hard glass-like surface. Special equipment must be installed to do this and it is only suitable for flat or gently curving surfaces. Very thick films (about $250 \mu m$) have to be applied to avoid sanding right through at edges and corners.

The other solution is chemical. Groups which dry by an oxidative mechanism are built into the polyester resin, so that oxygen causes surface hardening, even though styrene polyperoxide is formed. Unsaturated fatty acid groups are not used, but instead the shorter *allyloxy group*: $CH_2{=}CH{-}CH_2{-}O{-}$. This is very similar to non-conjugated unsaturation, because it contains a methylene group with a double bond on one side and an oxygen atom on the other. The oxygen atom is like a double bond in that both are strongly electron-attracting. The methylene group is very prone to oxidation by oxygen, hydroperoxide forms and oxidative drying proceeds as in Chapter 12. In addition, the allyl $(CH{=}CH{-}CH_2{-})$ double bond will polymerize slowly with the other ingredients by addition polymerization. The allyloxy group can be built into the polyester chain by using, as part of the original glycol, an allyl ether of a polyol with three or more hydroxyl groups, e.g.:

$$
\begin{array}{ll}
\text{CH}_2\cdot\text{O}\cdot\text{CH}_2\cdot\text{CH}{=}\text{CH}_2 & \qquad\qquad \text{CH}_2\cdot\text{O}\cdot\text{CH}_2\cdot\text{CH}{=}\text{CH}_2 \\
\mid & \qquad\qquad\qquad\qquad\qquad\quad\mid \\
\text{CHOH} & \qquad \text{CH}_3\cdot\text{CH}_2\cdot\text{C}\cdot\text{CH}_2\text{OH} \\
\mid & \qquad\qquad\qquad\qquad\qquad\quad\mid \\
\text{CH}_2\text{OH} & \qquad\qquad\qquad\qquad \text{CH}_2\text{OH}
\end{array}
$$

        glycerol monoallyl ether        trimethylol propane monoallyl ether

The disadvantage of this solution lies in the long surface hardening time (16–24 hours at room temperature; much less if heat is used and promoters, such as ethyl acetoacetate, $CH_3\cdot CO\cdot CH_2\cdot CO\cdot O\cdot C_2H_5$, are included to enhance the action of the cobalt). The allyloxy components also raise the cost of the finish. If a glass-like surface is required on verticals, sanding and polishing may still be needed, since perfect flow is difficult to achieve. On the other hand, where the surface is too complex to be sanded mechanically, this approach provides the only means of obtaining a glossy polyester finish.

## Radiation curing: polyesters and acrylics

On p. 215, ultra-violet light was listed as a cause of free radical formation. Polyester clear coatings stored in untinted glass bottles will polymerize in days or weeks. This process can be considerably accelerated and used as a method of curing the coating. This can be done by including in the coating a substance which is thermally stable, but which is readily decomposed to free radicals by u.v. light. Examples of such substances, which are called photo-initiators, are given below:

        benzoin ethers        benzil dimethyl ketal

benzophenone
(used with tertiary amines)

Very powerful u.v. lamps are available in tube form, rated at 80 watts per centimetre of lamp length or higher, and these can be mounted over a conveyor on which flat, coated articles (e.g. wood-based panels, metal sheet, paper or card) are carried. Hardening of the coating can be achieved by a few seconds of exposure to such lamps, because of the very efficient conversion of u.v. energy to free radicals. Yet the coating, being thermally stable, is a one pack product.

Even more rapid curing can be achieved if unsaturated polyester coatings are replaced by the more expensive *unsaturated acrylic coatings*. Instead of a polyester resin, a low molecular weight acrylic polymer is used, often called an *oligomer*. These acrylic oligomers are the reaction products of acrylic acid and end groups in epoxy resins and saturated polyesters, or of hydroxy monomers, such as hydroxyethyl acrylate, and polyisocyanates, e.g.

$$OCN\overset{\overset{\displaystyle NCO}{|}}{\rule{1cm}{0.4pt}}NCO + 3CH_2{=}CH{\cdot}CO{\cdot}O{\cdot}C_2H_4{\cdot}OH \longrightarrow$$

hydroxyethylacrylate

$$CH_2{=}CH{\cdot}CO{\cdot}O{\cdot}C_2H_4{\cdot}O{\cdot}CO{\cdot}NH\overset{\overset{\displaystyle NH{\cdot}CO{\cdot}O{\cdot}C_2H_4{\cdot}O{\cdot}CO{\cdot}CH{=}CH_2}{|}}{\rule{1cm}{0.4pt}}NH{\cdot}CO{\cdot}O{\cdot}C_2H_4{\cdot}O{\cdot}CO{\cdot}CH{=}CH_2$$

a urethane acrylate oligomer

These oligomers are dissolved in acrylic monomers. Suitable ones have low volatility, low odour and are not severe skin irritants. Generally these are di- or tri-acrylate esters, e.g.

$$\begin{array}{l} CH_2{\cdot}CH_2{\cdot}CH_2{\cdot}O{\cdot}CO{\cdot}CH{=}CH_2 \\ | \\ CH_2{\cdot}CH_2{\cdot}CH_2{\cdot}O{\cdot}CO{\cdot}CH{=}CH_2 \end{array}$$

1,6 hexane diol diacrylate

$$\begin{array}{l} \phantom{CH_3{\cdot}CH_2{\cdot}C}CH_2{\cdot}O{\cdot}CO{\cdot}CH{=}CH_2 \\ \phantom{CH_3{\cdot}CH_2{\cdot}C}| \\ CH_3{\cdot}CH_2{\cdot}C{-}CH_2{\cdot}O{\cdot}CO{\cdot}CH{=}CH_2 \\ \phantom{CH_3{\cdot}CH_2{\cdot}C}| \\ \phantom{CH_3{\cdot}CH_2{\cdot}C}CH_2{\cdot}O{\cdot}CO{\cdot}CH{=}CH_2 \end{array}$$

trimethylol propane triacrylate

Methacrylate oligomers and monomers can also be used, but rates of cure decline in the order

acrylates > methacrylates > polyester/styrene

Acrylate coatings are generally harder than polyester coatings and can be tougher or more flexible in thin films. Ingredients can be chosen to be substantially involatile at room temperature, while styrene is similar in volatility to xylene. Acrylates and methacrylates are, however, just as prone to air-inhibition as polyester/styrene systems.

In u.v. coatings, air-inhibition is usually overcome by creating so many free radicals in such a short time that the oxygen is 'swamped': although polyperoxides are formed, many successful carbon–carbon chain propagations also can occur. This kind of 'overkill' in the chemistry of the coating can be expensive and equipment has been developed to cure coatings with u.v. under a blanket of nitrogen. The complexities and costs of such equipment have not met with general favour, but the use of an inert gas blanket is accepted as an integral part of an alternative form of radiation curing.

This alternative is *electron beam curing*. In this technique, flat paint films are passed under an electron source, so that energetic electrons impact with the coating over its full width. Coating compositions are very similar to those used with u.v. sources (the order of reactivity of the chemical types is the same), but they do not contain free radical initiators of any type. Instead, the bombardment of high energy electrons creates free radical sites at random in the coating and polymerization is so rapid that cure takes place in a fraction of a second, provided that oxygen is excluded by a nitrogen blanket. The electron accelerators are very expensive and only those paint users with a very high output of flat painted objects (e.g. large manufacturers of flush doors) have so far been able to justify the expense.

While electron beam curing is equally effective with clear and pigmented coatings, u.v. curing has largely been confined to clear coatings. The problem is that most pigments absorb ultra-violet light quite strongly and prevent decomposition of photo-initiators in the depth of the film. Some success has been achieved with photo-initiators which absorb light in the very near u.v. (wavelengths above 350 nm) and with inks which, though heavily pigmented, are applied in very thin layers (2–3 $\mu$m).

## General properties of polyester finishes

Polyester finishes are very hard, tough, resistant to solvents and resistant to moderately hot objects (e.g. a teapot or a lighted cigarette). These are extremely useful properties for a furniture finish, clear or pigmented.

It is necessary to search among the other properties for the reasons why these finishes have found few other outlets.

Adhesion to most surfaces is poor, so that it is often necessary to use polyurethane and other special fillers and sealers, especially when there is no contact process basecoat. The appearance of good adhesion is assisted by the extreme difficulty of breaking fairly thick polyester films, in order to remove them.

Polyester films are not flexible, in the ordinary sense, unless soft. Sufficient flexibility to withstand the expansion and contraction of wood substrates as the humidity varies, is obtained, with hardness, by spacing out the cross-links. This can be done by decreasing the proportion of unsaturated acid in the polyester resin, or by using long-chain, flexible saturated acids or alcohols. The mechanical sanding and polishing properties of the finish will suffer if cross-linking is excessively reduced, since the finish will soften at the temperatures reached in these mechanical treatments (around 100 °C) and will be marked. Excessive clogging of sanding papers will also occur.

Unlike epoxy finishes, polyesters undergo appreciable volume shrinkage on hardening. This can cause loss of adhesion, pulling away from edges and other defects.

The ordinary maleic/phthalic/propylene glycol wood finish compositions have poor resistance to alkali, since the ester linkages are easily saponified, but more resistant polymers can be made at higher cost from *iso*-phthalic acid and from the reaction product of propylene oxide and bisphenol 'A':

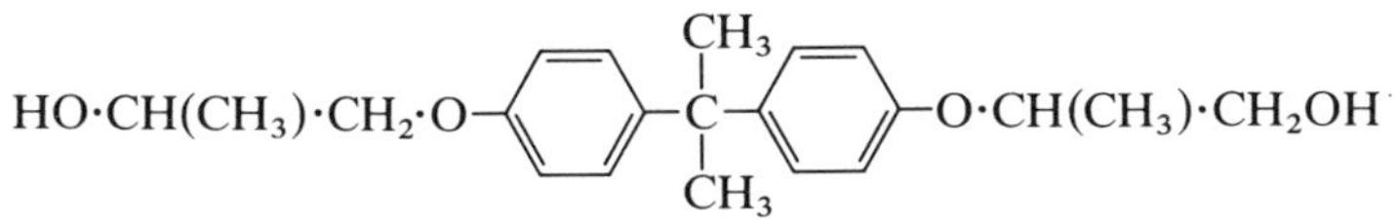

$$HO \cdot CH(CH_3) \cdot CH_2 \cdot O - \!\!\!\left\langle \bigcirc \right\rangle\!\!\! - \!\!\underset{\underset{CH_3}{|}}{\overset{\overset{CH_3}{|}}{C}} \!\!- \!\!\!\left\langle \bigcirc \right\rangle\!\!\! - O \cdot CH(CH_3) \cdot CH_2OH$$

Out of doors the varnish (wax type) has good gloss retention, but tends to fail by cracking, especially on expanding and contracting substrates. The pigmented types are usually based on allyloxy polyesters and they lose gloss rather quickly.

Examples of unsaturated polyester and acrylic coatings are given overleaf:

Clear wood finish (with wax)

| (a) Varnish | | wt % | |
|---|---|---|---|
| **Thickener** | Fine particle silica | 1·5 | disperse |
| **Resin solution** | Polyester resin (maleic/phthalic/ propylene glycol) at 67% solids in styrene | 35·8 | on roller mill |
| | Polyester resin solution (as above) | 40·7 | |
| **Monomer** | Styrene | 15·2 | |
| **Additives** | Wax solution in styrene (2% wax) | 5·0 | |
| | Cobalt naphthenate (6% Co) | 1·2 | |
| | 2% Hydroquinone in acetone | 0·6 | |
| | | 100·0 | |

| (b) Activator | | |
|---|---|---|
| **Peroxide** | 50% Methyl ethyl ketone peroxide in dibutyl phthalate | 25 |
| **Solvent** | Ethyl acetate | 75 |
| | | 100 |

Mixing ratio (for normal spray application, pot-mix or dual feed):
Varnish/Activator, 10/1 by volume
Pot-life: 10 mins. ⎱ 15–20 °C
Fully hardened: 4 hours ⎰

Alternatively, the finish may be applied over the following basecoat:

*Basecoat*

| **Resin** | HX30–50 (or RS ½ sec) nitrocellulose (dry weight) | 4·5 |
|---|---|---|
| | Cyclohexanone resin* | 4·5 |
| **Peroxide** | Cyclohexanone peroxide powder | 6·0 |
| **Solvents** | Methyl isobutyl ketone | 17·0 |
| | Acetone | 50·0 |
| | Ethyl acetate | 16·0 |
| | *Iso*propyl alcohol | 2·0 |
| | | 100·0 |

Spraying solids: 15%

The polyester finish should be thinned 12 : 1 by volume (varnish : thinner) with a 20 per cent solution of dibutyl phthalate in methanol.

* Cyclohexanone resin is made by polymerizing the ketone, with or without aldehyde, by heating under pressure with alkali.

Low gloss u.v. curing overprint varnish for paper or card

|  |  | wt % |
|---|---|---|
| **Matting aid** | Silica | 9·00 |
| **Oligomer solution** | Ebecryl 608 | 27·05 |
| **Monomer** | OTA 480 | 45·05 |
| **Photo-initiator** | ⌠ Benzophenone | 4·50 |
|  | ⌡ Uvecryl P101 | 13·50 |
| **Levelling aid** | Silicone oil | 0·90 |
|  |  | 100·00 |

Apply by roller coater at 100% solids.
Cure under one u.v. lamp (80 W/cm of length) at conveyor speed 20–40 m/min.
The following materials in the formula are made by U.C.B., s.a.:
*Ebecryl 608* is an epoxy-acrylate at 75% in OTA 480.
*OTA 480* is a polyol triacrylate.
*Uvecryl P101* is an unsaturated tertiary amine (with benzophenone it provides the photo-initiator system and is also a copolymerizable monomer).

# Seventeen

# Chemical treatment of substrates

In Part Two of this book so far we have considered the general principles on which all paints are formulated, the non-resinous ingredients that are used and, finally, six groups of paints classified by their drying mechanisms.

All these coatings are applied to a substrate to produce a paint system, e.g. substrate, primer, undercoat, topcoat. Whether the overall result is satisfactory will depend on how these layers adhere together, how they perform individually and where the weakest link is.

Sometimes the weakest link is at the substrate–primer interface (e.g. poor adhesion), or in the substrate itself (e.g. wood suffering from wet rot). For this reason, it is necessary now to consider the substrate itself and its particular chemistry.

Before we can paint the substrate to produce a successful system, it may be necessary to treat it in some way to upgrade its performance in that system. Such treatments are designed either (a) to improve the substrate, or (b) to prevent its degradation.

There are numerous reasons why it may be necessary to improve the substrate. It may be poorly produced, e.g. too rough to finish. It may have a natural weakness, e.g. chipboard is too porous. It may have been damaged or contaminated, e.g. dented or greasy metal. It may have deteriorated during storage, e.g. rusty steel.

Many of these problems can be dealt with by physically treating the substrate to produce a clean, level, sound surface suitable for painting. Such treatments include the planing and sanding of wood, the degreasing and sand-blasting of steel, and the filling and sealing of damaged or porous surfaces. Because they are essentially physical in nature, these treatments will not be discussed further in this Chapter. We shall confine our attention to those cases in which a chemical treatment of the substrate is necessary, either for improvement of it, or more often to prevent its degradation.

In particular, we shall look at the improvement of plastic surfaces for better paint adhesion, the protection of wood against biodegradation and the protection of metals against corrosion.

## Improving plastics for paint adhesion

Different plastics present different painting problems, but on the whole, once clean they do not often present problems of paint adhesion. Exceptions are the polyolefin plastics, e.g. polyethylene and polypropylene. Because these polymers lack polar groups, they have low energy surfaces which are difficult to wet and are not readily penetrated by solvents. In addition to this, the surface layer of the plastic, a few nanometres thick, is often different to the bulk, being low in molecular weight and weak. Paint adhering to this layer will pull the layer away and peel off, thus appearing to have poor adhesion.

To get good results on these plastics, it is necessary to modify the surface of the plastic, to crosslink the weak layer into the plastic below and to oxidize it, thus creating polar groups and a higher energy surface. The preferred treatments for doing this are corona discharge (for plastic films) and flame treatment (for moulded articles).

In *corona discharge* the plastic film passes at $1\text{–}2\,ms^{-1}$ over an insulated metal roller, which is electrically earthed and 1–2 mm under a bar electrode. Through the electrode passes an AC current oscillating at high frequency (about 20 000 Hz) and at a high voltage (about 20 000 V). From this electrode there is a continuous discharge towards the plastic film and earthed roller. Sparks can be seen, but additionally a *plasma* is produced. A plasma is a gas (in this case, air) in which the gas molecules are broken down into individual atoms, free radicals, ions, electrons and photons (in this case, by the high energy discharge).

These highly reactive particles impact with the plastic surface, oxidizing it and disrupting C—H and C—C bonds to a depth of 5–50 nm. Subsequent recombinations of the C· free radicals can cross-link the weak surface layer into the bulk polymer and the new surface will contain polar groups, such as $C{=}O$, C—OH, COOH, C—ONO and C—O—$NO_2$. These changes are reflected in a rise in the critical surface tension of the plastic from e.g. 31 to 47 dynes $cm^{-1}$. This surface is more readily wetted by paints. Adhesion is improved because surface layers are no longer weak and because of strong attractions between polar groups in paint and plastic (see pp. 89, 90).

In *flame treatment*, the oxidizing portion of a gas flame (6–10 mm from the tip of the blue inner cone) is brought into contact with the surface of the plastic for a brief period (0·02–0·1 s). The flame temperature is high (1 100–2 800 °C) and the flame is a plasma. Thus the plastic is modified by oxidation by the plasma and by recombination of free radicals created in the polymer with free radicals in the plasma. Changes are observed to a depth of 4–9 nm and wetting and adhesion are improved. This is a popular technique for improving the adhesion of printing inks to the surfaces of moulded polyolefin containers.

The above techniques are generally preferred, but any means of surface oxidation can be effective, such as treatment with oxidizing acids (strong sulphuric, chromic) or halogens, or treatment with photoinitiators (p. 222) followed by irradiation with UV light.

## Protection of wood against biodegradation

All surfaces are continually exposed to bacteria and the spores of *fungi*. The latter are particularly destructive to wood. The fungi which cause the most damage to wood are the members of the basidiomycetes family, which includes both brown and white rots. Brown rot is so called because it attacks only the cellulosic material in wood, leaving behind the lignin which is brown in colour. White rots, on the other hand, destroy both cellulose and lignin. Commonly encountered members of this family include *Serpula lacrymans* (a brown rot) commonly known as 'dry rot' and the wet rots *Coniophora puteana* (brown rot) and *Phellinus contiguous* (a white rot). Two other families are also important in the degradation of wood: these are the ascomycetes and the deuteromycetes (collectively known as soft rots). Apart from those fungi which cause structural degradation, many other fungi cause the surface disfigurement known as bluestain or sapstain. This drastically reduces the value of timber.

Perfectly sound-looking wood is almost sure to have spores on its surface and there may even have been penetration into the cells below. Even if the painter is very observant, he may well miss signs of degradation already present in the wood surface and covering organisms with paint will not prevent trouble later. This is especially so if design deficiencies in the wood structure aid penetration of water, encouraging the moisture content of the wood to rise above 20%, the level necessary for the fungi which cause rot to become active. They will then grow and multiply under the coating, feeding on the wood beneath and causing rot damage, or erupting through the coating to form pinhole damage and cause staining.

However, fungi can be prevented from attacking the wood by treating it chemically and keeping its moisture content to a minimum. Chemical treatment may be solvent-borne or water-borne. Active ingredients include pentachlorophenol, tributyltin oxide and copper and zinc naphthenates (solvent-borne) and mixtures of copper sulphate, sodium or potassium dichromate, and hydrated arsenic pentoxide (water-borne). These act in a variety of ways to destroy or inhibit the growth of fungi.

It is equally important to get these preservatives deep into the wood. For best results special equipment is required and the job is best done in the factory after seasoned pieces have completed all stages of machining. The most effective techniques use vacuum and/or pressure to displace air from the cells and force preservative into them. If remedial action has to be taken outside a factory, pressure injection, dipping, brushing and spraying may be used (in order of decreasing penetration).

Special low viscosity preservative-containing primers can also add markedly to the life of an exterior paint system by destroying organisms on or in the surface layers of the wood and also by penetrating open joints to seal end-grain, thus helping to keep the moisture content of the wood as low as possible.

## Protection of metals against corrosion

Corrosion is essentially the conversion of metal to a hydrated form of oxide. In the presence of oxygen and water, the overall corrosion reaction of steel is

$$4Fe + 3O_2 + 2H_2O \longrightarrow 2Fe_2O_3 \cdot H_2O$$

On pages 17 and 18, electrolysis at inert electrodes was discussed. If the electrodes are not inert, then at the anode metal dissolution occurs and metal cations are formed, e.g.

$$Zn - 2 \text{ electrons} \longrightarrow Zn^{2+}$$

The surface of steel is never uniform and, if it is in contact with a thin layer of aqueous electrolyte, then small electrical imbalances from site to site will lead to the formation of an electrolytic cell. At anodic sites, the *anodic reaction*

$$4Fe - 8e \longrightarrow 4Fe^{2+} \text{ (oxidation)}$$

occurs. The cations formed migrate through the electrolyte towards more negative, cathodic sites. At these, the *cathodic reaction*

$$4H_2O + 2O_2 + 8e \longrightarrow 8OH^- \text{ (reduction)}$$

is occurring and the anions produced migrate from thence towards the anodic sites. Within the metal, electrons flow from anodic to cathodic sites and within the electrolyte, migrating ions meet to form soluble ferrous hydroxide:

$$4Fe^{2+} + 8OH^- \longrightarrow 4Fe(OH)_2$$

When sufficient oxygen is present, this is oxidized to insoluble hydrated ferric oxide

$$4Fe(OH)_2 + O_2 \longrightarrow 2Fe_2O_3 \cdot H_2O + 2H_2O.$$

thus it can be seen that corrosion (dissolution of the metal) is an electrochemical process. Rusting does, however, require an additional oxidation step. If the final corrosion product can be formed as a strongly adherent, insoluble and impermeable layer on the metal surface, corrosion will decline, but if it is loose or permeable, the corrosion process will continue. In real life, the above processes are modified by the presence of other chemical species, even in trace amounts, and on steel the layer is commonly loose and permeable.

In fact, corrosion can be inhibited by one of the following techniques:

(1) Keeping the surface completely dry, so that no conducting electrolyte can form on it.
(2) Starving cathodic areas of oxygen.
(3) Forming a film impermeable to electrons on cathodic sites.
(4) Forming a film impermeable to metal cations on anodic sites.

Coating metal with paint might be assumed to be taking preventive action by all four mechanisms, but in fact no paint is *completely* impermeable to water or oxygen, so that after an initial delay, barrier properties are lost and corrosion eventually begins. It is therefore necessary either to include inhibitive ingredients within the paint (anti-corrosive pigments) or to apply an inorganic chemical pretreatment before painting, or to do both.

The objectives of using inorganic chemical pretreatments, often called *conversion coatings,* are two-fold:

(1) To *passivate* the surface by forming upon it a relatively stable, strongly adherent, corrosion-inhibiting layer.
(2) To provide a surface to which paint coatings readily adhere.

The conversion coatings are thin and easily damaged. Subsequent layers of paint protect the films from damage and, if they contain inhibiting pigments, repair faults and damage where they occur. The overall result of paint plus pretreatment is often prevention of corrosion for years.

The use of inorganic chemical pretreatments on metal will now be illustrated by considering the chromating of aluminium and the phosphating of steel.

## Chromating of aluminium

Aluminium alloys are widely used in aircraft construction, extrusions for building purposes, skins of caravans (aluminium coil) and beverage cans. In each case they are protected by organic surface coatings on interior and exterior surfaces. Prior to coating, the metal is pretreated to improve corrosion protection, normally with a chromate treatment. The most widely used of the well-established chromating treatments are the amorphous chromate process and the amorphous chrome phosphate process.

*Amorphous chromate treatments* contain hexavalent chromium (chromates, $CrO_4^{2-}$, and dichromates, $Cr_2O_7^{2-}$) and fluoride ($F^-$) and may contain an accelerator to speed up the reaction. They are strongly acidic (pH 1.5–2) and operate at temperatures between 20 and 40 °C. These treatments are preferred for aircraft components and coil-coated aluminium and produce a golden yellow coloured coating. Coating weights vary between 100 and 600 mg m$^{-2}$.

*Amorphous chrome phosphate treatments* contain hexavalent chromium, fluoride and phosphate and operate at about pH 1 and 40–50 °C. They are preferred for can coatings (especially can end stock) and give colourless to green films with coat weights of 50–500 mg m$^{-2}$.

Not only are there a wide variety of commercial chromate formulations in use, but the conversion coatings formed also differ in composition and experts disagree about the detailed chemistry. There is, however, general agreement that in amorphous chromate treatments the following reactions are involved:

(1) Acid attack upon the metal

$$2Al + 6H^+ \longrightarrow 2Al^{3+} + 3H_2$$

(2) Reduction of hexavalent chromium to trivalent chromium

$$3H_2 + 2Cr_2O_7^{2-} \longrightarrow 2Cr(OH)_3 + 2CrO_4^{2-}$$

(3) Formation of hydrated aluminium oxide

$$2Al^{3+} + 4H_2O \longrightarrow Al_2O_3 \cdot H_2O + 6H^+$$

(4) Formation of aluminium chromate

$$2Al^{3+} + 3CrO_4^{2-} \longrightarrow Al_2(CrO_4)_3$$

(5) Formation of complex chromic chromates

$$2Cr(OH)_3 + CrO_4^{2-} \longrightarrow Cr(OH)_3 \cdot Cr(OH)CrO_4 + 2OH^-$$

The role of the fluoride is to assist dissolution of the aluminium, even when in oxide form.

The amorphous chrome phosphate process forms coatings consisting largely of aluminium and chromium phosphates (see *phosphating* below). It is preferred by the can industry, because only trivalent chromium is found in the formed coating. In view of growing concerns about the toxicity of hexavalent chromium, this is an essential feature for containers for foodstuffs.

Concerns about toxicity and effluent disposal have led to other approaches for the pretreatment of aluminium. *No-rinse pretreatments* contain hexa- and trivalent chromium with organic polymer or silica and are applied at controlled film weights with no subsequent removal of excess by rinsing. They therefore present no effluent disposal problem. *Chrome-free pretreatments* (essentially special phosphating treatments) avoid chromium in the workplace altogether and have been widely adopted for two-piece beer and beverage cans.

## Phosphating of steel

The main phosphate pretreatments for steel are iron phosphate and zinc phosphate types. Zinc phosphates are preferred for best corrosion resistance out of doors.

*Iron phosphate* coatings are amorphous coatings and very thin (0.1–1 $gm^{-2}$). The working solution contains primary sodium or ammonium phosphates, together with other ingredients which may include an oxidizing accelerator, and surfactants to combine degreasing with the chemical treatment. The pretreatments operate between pH 3 and pH 5.5 at temperatures between 50 and 80 °C. The process begins with attack on the steel, forming primary ferrous phosphate:

$$8Fe + 16NaH_2PO_4 + 8H_2O + 4O_2 \longrightarrow 8Fe(H_2PO_4)_2 + 16NaOH$$

About a half of this is oxidized to ferric phosphate, which is insoluble and deposits:

$$4Fe(H_2PO_4)_2 + 4NaOH + O_2 \longrightarrow 4FePO_4 \downarrow + 4NaH_2PO_4 + 6H_2O$$

The other half of the ferrous phosphate is converted to ferric hydroxide as the pH rises:

$$4Fe(H_2PO_4)_2 + 12NaOH + O_2 \longrightarrow$$
$$4Fe(OH)_3 \downarrow + 4NaH_2PO_4 + 4Na_2HPO_4 + 2H_2O$$

and on drying:

$$4Fe(OH)_3 \longrightarrow 2Fe_2O_3 + 6H_2O$$

The final coating contains about 40% ferric oxide and the rest ferric phosphate.

*Zinc phosphate* coatings are crystalline and somewhat thicker (1–5 $gm^{-2}$). The bath or spray contains primary zinc phosphate, $Zn(H_2PO_4)_2$, phosphoric acid and oxidizing accelerators. The pretreatments operate between pH 1.5 and pH 3.3 at temperatures between 25 and 90 °C. The process begins with acid attack on the steel, oxidation (assisted by the accelerator) and precipitation of some ferric phosphate as above. However, the zinc phosphate is in a finely balanced equilibrium with the other species:

$$4Zn^{2+} + 3H_2PO_4^- \rightleftharpoons ZnHPO_4 + Zn_3(PO_4)_2 \downarrow + 5H^+$$

The removal of even small amounts of hydrogen ion by acid attack on the steel

$$Fe + 2H^+ \longrightarrow Fe^{2+} + H_2$$

upsets the equilibrium locally. As a result, the reaction moves to the right to create more hydrogen ion, producing as it does so sparingly soluble secondary zinc phosphate and insoluble tertiary zinc phosphate. The latter deposits at the sites where reaction has taken place, forming a protective crystalline layer, mainly hopeite, $Zn_3(PO_4)_2 \cdot 4H_2O$, though some phosphophyllite, $Zn_2Fe(PO_4)_2$, is also formed in the layers nearer the substrate.

Both iron and zinc phosphate coatings contain small amounts of porosity, probably caused by the escape of hydrogen in the earlier stages of phosphating. The free pore area can be reduced by rinsing in hot dilute (0.01–

0.05%) solutions of chromic acid. Insoluble iron chromates are formed in the pores, passivating these areas and increasing the corrosion resistance almost fourfold. Thus the protective properties of phosphating and chromating are combined. However, in many cases the phosphate pre-treatment is designed to give good results without the chromate rinse, either for reasons of economy, or because of concerns about the toxicity of hexa-valent chromium and the disposal of the effluent.

Iron phosphating is widely used on panel radiators, refrigerators and other items not subject to severe exterior exposure. Zinc phosphating is used to protect motor car bodies and other articles which require maximum corrosion protection. Since phosphating works well on zinc, it is used on electro-zinc for washing machines and on hot-dipped galvanized steel coil-coated for the cladding and roofing of buildings. The use of zinc-coated steels in automobile bodies is also increasing rapidly, so that phosphate pretreatments have to be designed to treat both zinc and steel surfaces.

# Appendix:
# suggestions for further reading

These publications cover the main paint ingredients and some special features referred to in Part Two of this book. The expensive reference books (marked*) can be consulted through technical libraries.

## Polymers

*The Chemistry of Organic Film Formers,* D. H. Solomon (Wiley, New York).

## Pigments

*Pigment Handbook Vol. 1: Properties and Economics,* edited by Temple C. Patton (Wiley, New York, 1973).
*Treatise on Coatings. Vol. 3: Pigments, Part I,* edited by R. R. Myers and J. S. Long (Marcel Dekker, New York).

## Assessment of pigment properties

\* *Paint Testing Manual,* edited by G. G. Sward, 13th edition (American Society for Testing and Materials, 1972). The most comprehensive volume on paint testing available.

## Pigment dispersion equipment

\* *Paint Flow and Pigment Dispersion,* T. C. Patton, 2nd edition (Wiley, New York, 1979) Chapters 17–26. A comprehensive book, which contains practical information on various mills, as well as a full theoretical coverage.

## Solvents

* *Solvents Guide,* C. Marsden and S. Mann (Cleaver-Hume).
* *Techniques of Chemistry. Vol. 2: Organic Solvents,* J. A. Riddick and W. B. Bunger, 3rd edition (Wiley, New York, 1970).

Both comprehensive reference books of solvent properties.

## Application of paint

*Paint Technology Manuals. Vol. IV: Application of Surface Coatings,* collated by C. W. Collier (Chapman and Hall, London, 1965).
*Industrial Paint Application,* W. H. Tatton and E. W. Drew, 2nd edition (Newnes–Butterworth, London, 1971).

## Paint testing

* *Paint Testing Manual,* edited by G. G. Sward, 13th edition (American Society for Testing and Materials, 1972).
*Paint Technology Manuals. Vol. V: The Testing of Paints,* collated by F. G. Dunkley and C. W. Collier (Chapman and Hall, London, 1965).

## Paint defects

* *Hess's Paint Film Defects,* edited by H. R. Hamburg and W. M. Morgans, 3rd edition (Chapman and Hall, London, 1979). A comprehensive reference book on all defects found in paint films, their causes and cure.

## Toxicity

* *Patty's Industrial Hygiene and Toxicology,* Vols. 2A, 2B and 2C, edited by G. D. and F. E. Clayton, 3rd edition (Wiley-Interscience, New York, 1981–82).
* *Dangerous Properties of Industrial Materials* N. I. Sax, 6th edition (Van Nostrand Reinhold, New York, 1984).
* *Hazardous Chemicals Data Book,* edited by G. Weiss, 2nd edition (Noyes, New Jersey, 1986).
*Hazards in the Chemical Laboratory,* edited by L. Bretherick, 3rd edition (Royal Society of Chemistry, London, 1981).
*Rapid Guide to Hazardous Chemicals in the Workplace,* edited by N. I. Sax and R. J. Lewis (Van Nostrand Reinhold, New York, 1986).

## General paint technology

* *Surface Coatings,* Vols. 1 and 2, Oil and Colour Chemists' Association, Australia (Chapman and Hall, London, 1983–84).
* *Paint and Surface Coatings: Theory and Practice,* edited by R. Lambourne (Ellis Horwood, Chichester, 1987).

# Index

The main reference is given in bold type where appropriate.

Abietic acid 162
Absolute zero 6
Absorption of light 76, **80–81**
Accelerator 135, 159–160
Acetaldehyde 38, **46–47**
Acetamide 48
Acetic acid 19, **38–40**, 46, 48–49
  anhydride **39–40**, 48, 49
Acetone 46, 48, 114
Acetyl chloride **39–40**, 48, 49
Acetylene 26, 35
Acid 17–21
  anhydride **39–40**, 48, 49, 189–190
  carboxylic 38–40
  -catalysed finishes **179**, 209
  chloride **39**, 49, 52
  fatty **39**, 41–43
  Lewis 190
  value(acid number) 21, 109, **167**
Acidity 20
Acrilan 67
Acrylamide 141, 177
Acrylic acid 141, 149, 188, 223
  coatings (unsaturated) 223–224
  finishes, thermosetting 178–181,
     187–188, 206, 223–224
  lacquers 140–143, 145
  latices 149–150, 179, 180
  nitrogen resins 176–177
  polymers 140–142
  resins, thermosetting **176–178**, 206
Activator 135
Active hydrogen **45**, 55
Acyl 39

Addition reactions 62, 172
Additive 87, 99, **125–137**, 209
Adhesion **89–90**, 183, 209, 228–230
Adipic acid 167
Aircraft finishes 193
Air-inhibition **220–222**, 224
Airless spraying 194
Alcohols 37–38
Aldehydes 46–47
Aldol 47
Algicide 136–137
Aliphatic 29
Alizarin 78
Alkali 19
Alkane 33
Alkoxy radicals 156
Alkyd finishes 169–170
Alkyds 142, **165–168**, 178–182, 205
  vinyl-modified 168
Alkyl 33
Allyl compounds 222
Allyloxy compounds 222
Aluminium alkoxide derivatives 160
  chloride 21
  metal 232–233
  oxide **21**, 44
  pigment 100, 150
  silicates 128
Amide group 48
  linkage 49
Amides 48–49
Amine-epoxy resin adduct 192
Amines **47–48**, 109, 179, 191–192,
    208, 209, 216

Amino groups  47
  resins  150, **171**
Ammonia  **18**, **27**, 45, 47–49, 109,
    110, 179
Ammonium acetate  48
  carbamate  49
  carbonate  26
  chloride  20, 27
  hydroxide  18, 20, 27
  ion  27
  nitrate  27
  nitrite  26
Amorphous solids  59
Amphoteric  21
Amyl acetate  40
*iso*-amyl acetate  40
Angle of incidence  71
  of reflection  71
Aniline  54
  hydrochloride  54
Anion  **17**, 18
Anode  17, 18
Anthracene  77
Anti-corrosive finishes  193
Anti-fouling additives  137
Anti-oxidants  136
Anti-skinning agents  136
Application  87–88
Aromatic compounds  50–54
Arsenic pentoxide  230
Atom  3
Atomic nucleus  3, 11, 13
  number  3
  weight  61
Attractions, inter-molecular  89–90
  inter-particle  99, 126–130
Attrition  105
Autodeposition process  150
Autophoretic process  150
Auxochromes  77–78
Azelaic acid  100
Azo group  77
Azomethine group  77
Azoxy group  77

Bactericide  147, 149
Barium bicarbonate  26
  peroxide  24

  sulphate  24
Base  17–21
Basecoat, active  219
Basicity  20
Bentones  129
Bentonite  128
Benzene  50–52
  sulphonic acid  50–51
Benzildimethyl ketal  222
Benzine  33
Benzoic acid  **53**, 168
Benzoin ethers  222
Benzophenone  223
Benzoyl peroxide  216–217
Benzyl alcohol  54
Bicarbonate  15, 19, 26
Binder  85
Biocide  **136**, 149
Biodegradation  230
Bisphenol A  **183–184**, 225
Bisulphate  19
Bisulphite  19
  compound  47
Bitumen  33
Bleeding  97, 102
Blushing  132
Boat varnish  200, 207
Boiling point  **9**, 22, 24, 112–115,
    **122–124**
  range, 112–115, **123**
Bond, chemical  11–13
  co-ordinate  11
  covalent  **11**, 24
  double  **35**, 60, 150–159, 161, 214, 222
  hydrogen  **22**, 58, 116, 118, 123, 130,
    131
  ionic  17
  length  51
  single  30
  triple  35
Boric oxide  21
Boron trifluoride  190, 205
Bromine  34
Bromo group  77
Brownian movement  10
1,3-butadiene  **35**
Butane  32–33
*iso*-butane  32–33

1,3-butane diol 205
1,4-butane diol 205
Butanol **37**, 113, 123, 173–177, 179
2-butoxy ethanol 111, 117, 179, 182
Butyl acetate 40
  acrylate 149
  alcohol **37**, 113, 123, 173–177, 179
*iso*-butyl alcohol **37**, 46, 175
*sec*-butyl alcohol 37
*ter*-butyl alcohol 37
Butyl benzyl phthalate 54, **142**, 145
  butyrate 40
*ter*-butyl perbenzoate 216
*iso*-butyl propionate 40
1-butylene 35
2-butylene 35
Butyric acid 39

Calcium
  bicarbonate 26
  carbide **26**, 175
  carbonate 15, **25–26**, 85
  chloride 13, 15, 25, 27
  cyanamide 175
  hydroxide 26
  oxide 26, 136, 175
  salts 159–160
Can finishes 152, 187, 191
ε-caprolactam 202
Carbamic acid 55
Carbides 26
Carbohydrate 139
Carbon 14, 23, **25–26**, 175
  black 98, 102
  dioxide 14–15, 23, **25–26**, 49, 202, 208
  monoxide 23, **25**, 26
  tetrachloride 35
Carbonate 15, 19, **26**
Carbonic acid 15, 19, 25
Carbonyl group **46–47**, 77, 178
Carboxyl group 38
Catalyst **24**, 135, 164
Cathode 17, 18
Cation **17**, 18
Cellobiose 139
Cellulose **139**, 230
  acetate 139
  acetate butyrate 140

colloids 131, 147, 149
  nitrate **140**, 142–143, 179, 226
  polymers 139–140
Cerium salts 159–160
Chain reaction 60–61
  transfer 61
  transfer agent 61
Chalking **91**, 185, 191, 193, 200
Charge transfer 78
Chelates, metal **131**, 150
Chloride 19
Chlorinated rubber 137, 138
Chlorine 23, 34–35, 45
Chloro group 77
Chloroethane 41
Chloroform 34
Chlorohydrin 45
Chloroparaffins 34
α-chloropropionic acid 38
Chroman ring 164
Chromates 19, 100, 232–233, 235
Chromating 232–233
Chromic acid **19**, 150, 230, 235
Chromophore 77–78
Cissing 133–134
Citraconic anhydride 214
Citric acid 19
Coalescing solvent **144**, 149
Cobalt salts 159–160, 216
Coefficient of viscosity 8
Co-grinding of pigments 108
Coil coatings **152**, 153, 178, 180, 187
Colloidal state 10, 129–130
Colloids, protective **147**, 148
Colour 70–81
  brightness 79–80
  chemistry 77–78
  Index 103
  matching 103, **107–108**
  mixtures 78–81
  pastel 96
  primary 79
  secondary 79
  solution 107
Combination finish 179
Complementary lights 79
  pigments 80
Compound 6

Condensation 9
  reaction 62–63, 172–173
Conjugation 42–43
Consistency 88
Contact process 219
Continuous phase 143
Contraction in cross-linking 194, 225
Conversion coatings 232–235
Co-ordinate bond 11
Copolymer block **62**, 130
  graft **147**, 108
  random 62
Copper 22, 27
  naphthenate 230
  nitrate 27
  oxide 22
  sulphate 78
Corona discharge 229
Corrosion 231–232
Cost of solvents 125
Coumarone 163
  –indene resins 163
Covalent bond 11
Cracking of paraffins 35
Critical surface tension **89**, 229
Crotonaldehyde 47
Crystallinity in polymers 65–69
Crystals 57–58
Cumene hydroperoxide 216, 217
Cupric chloride 78
Cuprous oxide 137
  thiocyanate 137
Curtain-coating 219
Cyanamide 175
Cyclohexane 50
Cyclohexanone peroxide 216, 217
  resin 226

Dammar 142
Defects, application 88
Diacyl peroxide 217
Diakon 141
Dialkyl peroxide 217
Diallyl phthalate 216
Diamond 25, 58, 73
Diazonium salt 190
Dibasic acid 39
Di-*t*-butyl peroxide 2

Dibutyl phthalate **142**, 148, 195, 218
2,4-dichlorobenzoyl peroxide 216
1,2-dichloroethylene 35
Dichromates 232, 233
Dicyandiamide 175, 189
Dielectric constant 18
Diels–Alder reaction 161
Diepoxide 0 **185**, 193
Diethylamine 203
Diethyl ether **44**, 115, 123
Diethylene glycol 111, **115**, 168
  monomethyl ether 111
  triamine 192
Diffraction 74–75
Dihydric alcohol 37
Diisocyanate, aliphatic 209–210
Dilatency 128
Diluents 124
  for epoxy resins 185
Dimethyl formamide 67
Dimethylamine 47
N,N-dimethylethanolamine **203**, 208
Dimethylol urea 172
Dimer fatty acids 192–193
Dipentene **36**, 111, 112
Diphenylmethane-4,4'-diisocyanate
  **55**, 204, 208
Dipping 87, 88
Diradical 155–156
Dispersed phase 143
Dispersion of pigments 103–107
  of polymer 148
Disproportionation 61
Dissociation 18–20
  degree of 19
Distemper, oil-bound 147
Dodecyl alcohol 134
Domestic equipment finishes 178,
  180, 188
Double decomposition 15
Driers 135, 154, **159–160**
Drying, lacquer 92, 139
  methods, comparison of 94–95
  of conjugated oils 154–156, 159
  of non-conjugated oils 154–155,
  156–159
  of paint 92–95
  oxidative 154–160

Drum finishes 187
Dual-feed polyester finishes 218–219
Durability 90, 162, 165, 169

Electrodeposition 110
  paints 186, 203
Electrolysis 17–18
Electrolyte 17
Electron **3**, 11–12, 13–14, 17–18, 21–22,
    24–25
  beam curing 224
  valency **11–13**, 14, 17, 21–22, 24–25
Electronegativity **13–14**, 17, 21–22,
    24–25
Element 3, 6
Eleostearic acid **42**, 156
Emulsification 146–147
Emulsion 11, 143–144, 146–152
  aqueous 110, **146–148**
  paint 143–153
    application 146
    aqueous 148–152
    film-formation 144, 146
    glossy 149
    industrial 150–152
    oil-based 147
    pigmentation 144, 146
    solid 149–150
  polymerization 147–148
Enamel 86
Epichlorhydrin 183–185
Epoxides 183–185
  of dihydric alcohols 183, 191
Epoxy-acrylic finishes 187–188
  resin 185
Epoxy-alkyds 186
Epoxy-amino resin finishes 187, 188
Epoxy-ester resins **185–186**, 191
Epoxy finishes 183–197
  non-stoving 191–197
  solventless 193–195, 197
  stoving 186–191, 195–196
Epoxy-M/F-alkyd finishes 188
Epoxy novolac resin **185**, 191
Epoxy-polyamide finishes, 192–193
Epoxy polyamine finishes 191–192, 197
  coal tar-modified 193
Epoxy-phenoic finishes 187

Epoxy properties 183–185
Epoxy-resins 180, **183–186**, 223
  *cyclo*-aliphatic **185**, 193
  modified **185**, 193
Equations, chemical 14–16
Ester gum 142
  linkage 40
Esters 40–41
Ethane **29**, 35, 51
Ethanol 37–38
Ethanolamine 45
Ethenyl group 77
Ether linkage **44**, 139, 156, 205
Ether-alcohols 45, **111**
Etherification 44, **173**, 176
Ethers 44
  Ethyl acetate 40, 49
  acetoacetate 222
  acrylate 141, 148, 177
  alcohol 37–38, 40–41, 44, 47
  bromide 48
  cellulose **139**
  chloride 41
  hexoic acid 159
  hydrogen sulphate 41
  nitrate 41
  orthoformate 202
  sulphuric acid 38, 41
Ethylamine 48
Ethylene **35**, 38, 45, 51, 60–61
  copolymer latices 149
  diamine 192
  dichloride 36
  glycol 35, **37**, 45, 110, 165, 167
    dimethylacrylate 216
    monoacetate 46
    monobutyl ether 111, **115**
    monomethacrylate 141
  oxide **45–46,** 48, 205
2-ethyl hexyl acrylate 148
Evaporation **9**, 88–92
  rate 122–124
Extender **86**, 98–100, 135

Ferric chloride 21, 23
  hydroxide 234
  oxide 13, 15, 21, 231, 234
  phosphate 234

Ferrous chloride 23
  hydroxide 231
  phosphate 234
  sulphide 14
Filler 86
Film properties 89–92
Film-former **85**, 86, 94
Finish 86
  for car repainting 206
  for chipboard 179
  for concrete 193, 208
  for hardboard 179, 206
  for plastics 138, 207
  for rubber 207
Flame treatment 229
Flammability 124
Flash point **112–115, 124**
Flexibility **90–91**, 206, 207
Flocculation **99**, 107, 127–130
Floor finish 193, 200, 202, 207, 211
Flow 88
  agents 134
  cup 118, 120
  time 120
Fluoride 232, 233
Force-drying 193
Formaldehyde **46**, 163–164, 171–177, 180
Formalin 172
Formic acid 39
Formula, acid-catalysed woodfinish 181
  acrylic metallic car finish 181
  alkyd gloss paint 169
  anodic electrodeposition primer 196
  can coating, interior 196
  chemical 6
  coil coating plastisol 151
  emulsion-based undercoat 151
  emulsion paint 151
  lacquer 145
  oil-modified polyurethane paint 211
  oleoresinous varnish 170
  paint (ball mill dispersion) 107
  polyester basecoat 226
  polyester woodfinish 226
  prepolymer finish 211
  solventless epoxy finish 197
  stoving enamel 181

urethane two-pack finish 212
  UV-curing varnish for paper 227
  water-reducible stoving finish 182
  wood primer 170
Free radicals **24**, 60–62, 155–160, 215–216, 219–221, 222–223, 224, 229
  combination of 61
Freezing point 9
Frequency 70
Fumaric acid 214
Functionality of alkyds 168
  of drying oils 156, 159
  of glycerol 167
  of monomers 63–65
Fungi 230
Fungicide 136, 147, 149
Furniture finishes 138, 145, 207, 219, 224–226

Gases 9
  elementary 14
Gas oil 33
Gel, irreversible 93
Gel strength, measurement 132
Giant molecule 58
Glass temperature **59**, 66
Gloss 72
  loss of 91
  reduction of 135
Glucose 139
Glycerin nitrate 41
Glycerine (glycerol) **37**, 41, 166–167
Glycerol monoallyl ether 222
Glycidyl methacrylate 185
Glycol ethers 45
Graphite 25
Grease 33
Grey 80

Hardness **90**, 206, 207, 224–225
  of water, temporary 26
Hemiacetal 47
Heterolytic dissociation 24
Hexamethoxymethyl melamine resins **176**, 179
Hexamethylene diisocyanate 55, 210
1,6-hexane diol 205

1,6-hexane diol diacrylate 223
1,2,6-hexane triol 205
Hiding power **97–98**, 128
High solids paint 153, 160, 176
  speed disperser 105
Homolytic dissociation 24
Homopolymer 62
Hot spray 143
House paints 148–151, 169–170
Hue **76**, 102
  brilliance of 102
Hydrocarbons, aliphatic **29–35**,
      111, 112
Hydrochloric acid 15, **18–21**, 25, 41,
      45, 47, 54
Hydrogen 12, 17, **22–24**, 27, 35
  bond **21–22**, 58, 116, 118, 123,
      130, 131
  iodide 44
  ion 18–20
  peroxide 24–25
  sulphide 19
Hydrogenated castor oil **43**, 129
Hydrolysis 40
Hydroperoxides 154–160, 216–217, 222
Hydroquinone 215
Hydroxide 15
  ion 18–20
Hydroxyethyl acrylate 223
Hydroxyethyl cellulose 149
Hydroxyl group **37**, 77

Incompatability 75
Indene 163
Indicator 20–21
Industrial equipment finishes 138, 196
Infra-red radiations **71**, 179
Inhibitor **136**, 215
Initiator 60–61
Interfacial tension 133–134
Iodo group 77
Iodonium salt 190
Ions **17–20**, 23
Iron 13–15
  oxides (pigments) 101
  salts 160
Isocyanate group 54–56
Isocyanate: hydroxyl ratio 207

Isocyanates **54–56**, 198–212
  blocked 202–203
Isocyanurates 204
Isomer **32**, 35, 51, 52–53
Isomers, geometrical 35
Isophorone diisocyanate 210
Isoprene 36
Itaconic acid 214

Jelly paint 127

Kerosene 33
Ketone peroxide 216–217
Ketones 46–47
Ketoxime **48**, 170

Lacquer 86, **138–143**
  film-formers 141–143
Lactic acid 38
Latent heat
  of evaporation 9
  of fusion 7
Latex (latices) 147–150
Laughing gas 13
Lauroyl peroxide 216–217
Lead 12
  carbonate 26
  chromate 100
  iodide 78
  monoxide 23, 26, 28
  nitrate 26, **28**
  red 100
  salts 100, 159
  sulphate **25**
  sulphide 25
Licanic acid 42
Ligand 78
Light 70–76
Lightfastness **97**, 102
Ligroin 33
Lime 26
Limonene 36
Linoleic acid 42
Linolenic acid 42
Liquids 7–9
Litharge 23
Litmus 17–20
Livering 100
Lucite 141

Magnesium 22, 27
  bicarbonate 26
  nitride 27
  oxide 22
Maintenance finishes 193, 200
Maleic acid 39
  anhydride 110, **214**
Manganese dioxide 23
  salts 159–160, 165
Marine finishes 200, 207
Melamine 175–177
  formaldehyde resins **175–176**,
    178–179
Melting 7
Metal finishes 138, 141, 165, 178,
    180–182, 187, 207
Metal primer 162, 187
Metallic elements **6**, 14
Meta-substitution 52
Methacylic acid **141**, 149, 177, 188
Methane **29–30**, 33
Methoxy group 77
Methyl alcohol (methanol) 37
  ethyl ketone 46, 114
    peroxide 216–217
  iodide 44
  isobutyl ketone 46, 114
  methacrylate **60**, 149, 177, 215
  propyl ether 44
Methylamine 47
  hydrochloride 47
Methylene chloride 34
  group 34
Methylol group 172
  ureas 172
Mica pigment 100
Micelle 148
Microgels 130–131
Micro-organisms 136–137
Mill 104
  ball 105
  bead 105
  high speed disperser 105
  roller 105–106
  triple-roll 106
Mixer, heavy-duty 106
  pug 106
Mixture 6

Molecular sieves **202**, 208
  weight **61**, 63
Molecule 6
Monoglyceride **166**, 198–199
Monomer **60–65**, 215–216
Monoterpene 36
Montmorillonite 128
Motor car finishes 138, 145, 178,
    181, 206
  primers 187
  surfacers 187

Naphthenic acids 159
Natural gas 33
  resins 142
Neutralization 20–21
Neutron **3**
Nitrate 15, 19, **28**
Nitric acid 19, 27, 34, 41, 53
  oxide 27
Nitrides 27
Nitrites 19, 26–27
Nitro group 77
Nitrocellulose **140**, 142–143, 179
Nitroethane 41
Nitrogen 12, 14, **26–28**, 166, 175, 224
  peroxide 27
  resins **171–175**, 205
    effects of alcohol 174–175
    paints based on 178–182
Nitroglycerine 41
Nitroparaffins 34
Nitrophenols 53
Nitroso group 77
Nitrous acid 19, **27**
  oxide 13, **27**
Noble gases 12
Nonanol 1
Non-aqueous dispersions 146, **147**, **148**,
    152, 178
Non-metals **6**, 14
Novolac, phenolic 164
Nucleus of atom **3**, 11, 13
Nylon 69

Occupational Exposure Limits **125**,
    180, 210
Octadecane 33

Octoic acid 159
Oil absorption 98–99
  blown 161
  bodied 161–162
  boiled 161
  castor **42–43**, 198, 202, 205, 208
  coconut 42
  conjugated **155–156**, 161
  dehydrated castor 43
  drying 41–42
  essential 36
  fatty 41
  heat-bodied 161
  hydrogenated castor 43
  length of alkyds 168
    of epoxy ester 186
    of oleoresinous varnish 165
    of urethane alkyds 199
    of urethane oils 199
  linseed **42**, 154, 162, 165
  lubrication 33
  maleinized **110**, 186
  non-conjugated 154–155, **156–159**
  non-drying 42–43
  oiticica 42
  paints 154–165
  perilla 42
  safflower 42
  semi-drying 42
  soya bean **42**, 168
  stand 161–162
  tall 42
  tobacco 42
  tung **42**, 154
Oils 41–43
Olefins 34–35
Oleic acid 42
Oleoresinous finishes **162–165**, 169–170
Oligomer, acrylic (unsaturated) 223
Organosols 68, 146, **152–153**
Orthophosphate 19
Orthophosphoric acid 19
Ortho-substitution 52
Oxidation **22–23**, 229–230, 231, 234
Oxides 21
Oxidizing agent **23**, 25, 27
Oxygen 12–15, 17–18, **22–25**, 154,
    161, 215, 221–222, 224

Paint 85–87
  literature 101, 103
Paper finishes 138
Paraffin oil 33
  wax **33**, 221
Paraffins 29–34
Paraformaldehyde (paraform) 172
Para-substitution 52
Particle shape 99–100
  size **98–99**, 101
Passivation 232
Patching 91–92
Pauling 14
Pentachlorphenol 230
Pentadecane 33
Pentaerythritol 37, 167, 198
Pentane 32–33
*cyclo*-pentane 32
*iso*-pentane 32–33
*neo*-pentane 32–33
*neo*-pentyl glycol **167**, 178
Peracid ester 217
Periodic table **3–6**, 11
Peroxide linkage 158, 216, 217
  radicals 155–160
Peroxides, organic 216–217
Perspex 60, 141
Petrol 33
  light 33
Petroleum 33, 34
  resins **162–163**, 165
  spirit 111
pH scale **20**, 149, 150, 232, 233, 234
Phenol 53, 164, 202
Phenolic condensate 163
  resins 162, **163–165**, 187, 196
Phenols **53**, 195, 215
Phenyl group 54
  isocyanate 54
Phosgene **49**, 54
Phosphates **20**, 232–235
Phosphoric acid 19, 20
Phosphorus oxychloride 39
  pentachloride 39
Photoinitiators 190–191, 222–223
*iso*-phthalic acid **53**, 167, 178, 214, 225
Phthalic anhydride **53**, 165–167, 200,
    214

Pigment 85, 96
  anti-corrosive **100**, 232
  chemistry 100
  dispersion 103–107
  functions 96
  literature 101, 103
  particle shape 99–100
  properties 96–101
  selection 103
  surface 99, 128
  surface area 98–99
  volume concentration 91, 128
Pigments, azo 102
  black 102
  dioxazine 102
  inorganic 101–102
  natural 101
  organic 102
  perinone 102
  perylene 102
  phthalocyanine 102
  quinacridone 102
  synthetic 101
  white 98, 100, 102
α-pinene 36
Pine oil 36, **111**, 112
Plasma 229
Plastic flow 127
Plasticisers 91, **142**, 148, 149, 151, 153
Plastics, surface treatment 229–230
Plastisol 68, 151, 153
Platinum 22, 27
Plexiglas 141
Poise 9
Poiseuille 9
Polarity 21–22
Polishing 91
  mechanical 221
Polyacrylonitrile 67
Polyamide resin 129, **192–193**, 194
Polyamines, aromatic 194
Polybasic acid 39
Polybutyl methacrylate 62, 141
Poly-*iso*-butyl methacrylate 141
Polycoumarone 163
Polyester finishes (unsaturated)
      213–226
  properties 224–225

resins 63, 166
  saturated **165–166**, 178, 180,
      205–206, 208, 223
  unsaturated **213–214**, 222, 225
Polyethers 110, 134, **205**, 208
Polyethylene 60, 67, 229
  glycol 110
Polyhydric alcohol **37**, 198
Polyisocyanurates 204
Polylauryl methacrylate 141
Polymer **59–69**, 85
  addition 60–62
  amorphous 66–67
  branched 64–65
  condensation **62–63**, 69
  cross-linked **64–65**, 90–94
  crystalline **65–69**, 139, 141
  dispersion 148
  linear **64–65**, 90–92
  molecular weight **61**, 63, 88, 90,
      93–95
  thermoplastic 65
  thermosetting 65
Polymerization, cationic 190–191
Polymethyl acrylate 148
Polymethyl methacrylate 60, 62,
      **141–143**
Polyol 37, **198**
Polyperoxide 156, 221
Polypropylene 229
Polystyrene 68–69
Polythene 60
Polyurethane finishes
  one-pack **198–203**, 211
  two-pack **203–208**, 212
  properties 208–210
Polyvinyl acetate 60, **68**, 148
  alcohol 131
  chloride 60, **68**, 151, 152, 153
Polyvinylidene fluoride 152–153
Popping 132
Potassium bicarbonate 26
  carbonate 26
  chlorate 23
  chloride 23
  hydroxide 19–20
  nitrate 28
  nitrite 28

permanganate 35, **38**
sulphate 19
Pot-life **94**, 208, 218–219
Pot-mix polyester finish 218
Powder coatings **188–190**, 202
Prepolymers **200–202**, 209
Pressure of a gas 9
Pretreatments 232–235
  amorphous chromate 232–233
  chrome-free 233
  chrome phosphate 232–233
  iron phosphate 234
  no rinse 233
  zinc phosphate 234–235
Primary alcohol 37
  carbon 34
Primer 86
  oil-based 162
  preservative 231
  –surfacer 86
Propane 29–31
Propanol 37, 44, **113**
*iso*-propanol 37, **113**, 123
Propionaldehyde 46
Propionic acid 38, 39
*iso*-propyl 33
Propyl alcohol **37**, 179
*sec*-propyl alcohol 37
Propylene 35, 45
  chlorohydrin 45
  glycol **167**, 214
    monomethyl ether 111
  oxide 45, **205**, 225
Proton **3**, 13, 17
Pseudoplasticity **127**, 128
Pulling over 91

Quaternary
  ammonium hydroxide 48
    ions 48, 128
    salt 48
  carbon 34
  ethyl ammonium bromide 48

Radiation curing 190–191, 222–224
Radiations, electromagnetic 70–71
Reactions, chemical 14–16
Reducing agent 23

Reduction 22–24
Reflection 71–72, 73–74
Reflow of lacquers 91
Refraction 72–74
Refractive index **73**, 75, 76, 98
Renovation 91–92
Repair 91–92
Resin **59–60**
  natural 59–60, 142
  water-soluble **109–110**, 176, 179,
    182, 186
Resistance to abrasion 188, 202, 209
  to alkalis 200
  to chemicals 138, 186–188, 200
  to solvents **91–92**, 203
  to water 99–100, 200, 205
Resole, phenolic 164
Retarder 135–136
Ricinoleic acid 42–43
Ring-opening 45–46
Road surfacing 208
Rosin 137, **162**, 164
Rust 13, **231**

Salt, common 6, **15**, 57
Salts 19–20
Sand-grinder 104–105
Sanding, mechanical 221
Sandwich process 219
Saponification 40
Satin sheen 135
Scavengers, moisture 136, **202**
Sealer 86
Sebacic acid 167
Secondary alcohol **37**, 46
  carbon 34
Second-order transition temperature 59
Shear 104–106
  rate of **8–9**, 126–127
  stress **8**, 126
Shelf-life 94
Shrivelling 160
Silica **128**, 135
Silicates **128–129**, 202
Silicone oils 133–134, 195
  resins **133–134**, 195
Silver 45
Sinkage 219

Skinning of paint 93
Soap formation 40
Soaps **39**, 40, 134, 147, 159
Sodium acetate **40**
  aluminate 21
  bicarbonate 26
  bisulphite 20, 47
  borate 21
  carbonate 25–26
  chloride 6, **15**
  hydroxide 15, **18**, 20, 25, 40, 45,
    184, 205
  nitrate 26
  phenoxide 52
  stearate 39
Softening temperature **59**, 66–68
Solids 7
  content **87**, 94
Solubility of crystalline solids 10
  of polymers 91, **116**, 118
  parameter **112–115, 116–119**, 121
Solubilization 148
Solute 10
Solution, saturated 10
Solutions 10
Sovency 116–118
Solvent properties 111–125
Solventless finish 193–195, 197, 208
Solvents 10, **109–125**, 129–132
  chemical types of 109–111
  for polyurethanes 209
Specific gravity of pigments 100
Spectrum, optical 76
  solvent 117, 118
Spraying 87–88
  twin-feed 194, 208, **218–219**
Stability of suspensions 103–104
  thermal, of pigments 101, 102
Stabilization, ionic 146–147
  steric 146–147
States of matter 6–9
Stopper 86
Strengths of acids and bases 19
Stress **8**, 126
Strontium bicarbonate 26
Styrene 53, 177, 215, 220–221
  –butadiene polymers 149
  copolymers 69

polyperoxides 221–222
Substitution of benzene 50–53
Substrate 86, **228–235**
Succinic acid 39, 167
Sulphate 15, 19–20
  ion 18
Sulphite 19
Sulphonium salt 190
Sulphur 12, 14, 23
  dioxide 22–23
  trioxide 18, 22
Sulphuric acid 19, 23, 24, 38, 41, 43,
    44, 50, 230
Sulphurous acid 19–20
Surface area **98–99**, 128
  optical 71
  tension **7**, 132–134
Surfacer 86
Surfactant **133–134**, 146–149
Suspension 10
  colloidal **10–11**, 103
Symbol 3, 6

Tank linings 194
TDI trimer 204
Terephthalic acid 167
Termination of polymerization 61
Terpene resins 163, 165
Terpenes **36**, 111, 112
$\alpha$-terpineol 36
Tertiary alcohol 37
  carbon 34
Terylene 69
Tetrahydric alcohol 37
Textile finishes 138
Thermoplastic 65
Thermosetting 65
Thickener 129–131
  selection 131–132
Thickeners
  associative **130**, 132
  mineral **128–129**, 132
  resinous **129–131**, 149
Thinner 87
Thio group 77
Thixotropy 127
Threshold limit values **125**, 180, 210
Tin salts 136, 137, 203, 209

Tinting strength 96–97
Titanium
  chelates 131
  dioxide 73, 98, 100, 102
Titration 20–21
Toluene 53, **113**
*p*-toluene sulphonic acid **179**, 181
Toluene diisocyanate 55, 201–202, 204
Topcoat 86
Toughness 90, 202, 209
Toxicity of formaldehyde 180
  of lead 159
  of polyurethanes 210–211
  of solvents 125
Transition elements 78
Triallyl cyanurate 216
Tributyl tin oxide 230
Tri (dimethylaminomethyl) phenol 195
Triethanolamine 45
Triglyceride 41
Trihydric alcohol 37
Trimethylamine 47
Trimethylol propane **167**, 204
  monoallyl ether 222
  triacrylate 223
Turpentine 36, 111, 112
Twin-feed spray guns 218–219
Two-pack paints **94**, 178, 179, 191–195,
    197, 203–208, 212, 213–226
Tyndall effect 11

Ultra-violet curing 190–191, 222–224
  lamps 223
  light **71**, 91, 97, 155, 169, 215, 230
Undercoats 86
Unsaturation 35
Uralkyd 199
Urea **49**, 58, 171–173
  formaldehyde resins 171–175
    water-soluble 172
  linkage **201**, 206–207, 210
Urethane 55
  alkyds **198–200**, 209
  oils **198–200**, 209

Valencies of carbon 29–30
Valency 11–13
  electrons **11**, 13–14, 17

variable **12–13**, 23
Vapour curing 208
Varnish **86**, 200, 202
Vaseline 33
Vehicle 86
Versatic acids 148
Vinyl
  acetate 60, 148
    copolymers 148
  chloride 60
    copolymers 68, 150
    latices 149, 150
  group 60
  monomer 215
toluene **53**, 215
  versatates 148
Vinylidene chloride copolymer
  latices 149, 150
Vinylidene group 60
  monomer 215
Viscometers 118, 120
Viscosity **8**, 88, 118, 120–122, 126–132
  brushing 169
  coefficient 8
  of emulsion 122
  measurement **118**, **119**, 126
  non-Newtonian 127–132
  of solutions 120–121
  of solvents 112–115, 121

Water 6, 12–13, **17–20**, 22, 109–111,
    112, 201–202, 206–208
  –based paints 109–110, 143–152, 176,
    179, 182, 195, 203
  –solubility of resins 109–110
Wavelength **70–71**, 74, 76
Wax 135
  paraffin **33**, 221
  polyethylene 135
  polypropylene 135
Weathering 91
Wet-on-wet polyester finish 219
Wetting **89–90**, 99, 103, 134
White spirit 111, 112
Whitewash 85
Whiting 85
Wire enamels 165
Wood primers 162, 170

Wood finishes 138, 145, 179, 181, 207, 213, 219, 223, 224–226
Wood rots 230

Xylene 53, **113**

Yellowing **165**, 169, 209–210

Zinc 24
 naphthenate 230
 oxide 100
 phosphate 233–235
 sulphate 24
Zirconium 159–160
 chelates 131
 complexes 169–170, 211